For thirty years Alan Blunt worked the woolsheds in outback Queensland, starting as a rouseabout and eventually working his way up to wool presser. He was a keen observer of character, and in the evenings recorded the portraits of workmates by the light of a pressure lamp, filling one foolscap pad after another. Later he had stories published in the outback and metropolitan press. In 1993 he formed a theatre troupe based at Longreach called Banjo's Outback Theatre & Woolshed, which he ran successfully for more than twenty years. Known as 'Banjo', he still performs one-man folklore and history shows, and makes guest appearances on ABC radio.

WOOL AWAY, BOY!

Alan Blunt

WILLIAM HEINEMANN: AUSTRALIA

Pseudonyms have been used in this book and other details altered where necessary to protect the identities and privacy of people mentioned.

A William Heinemann book
Published by Penguin Random House Australia Pty Ltd
Level 3, 100 Pacific Highway, North Sydney NSW 2060
www.penguin.com.au

First published by William Heinemann in 2016

Copyright © Alan Blunt, 2016

The moral right of the author has been asserted.

All rights reserved. No part of this book may be reproduced or transmitted by any person or entity, including internet search engines or retailers, in any form or by any means, electronic or mechanical, including photocopying (except under the statutory exceptions provisions of the Australian *Copyright Act 1968*), recording, scanning or by any information storage and retrieval system without the prior written permission of Random House Australia.

Addresses for the Penguin Random House group of companies can be found at global.penguinrandomhouse.com/offices.

National Library of Australia
Cataloguing-in-Publication entry

Blunt, Alan, author
Wool away, boy!: a ripping memoir of life in the shearing sheds/Alan Blunt

ISBN 978 0 14378 036 6 (paperback)

Blunt, Alan
Sheep shearers (Persons) – Biography
Sheep-shearing – Australia
Shearing sheds – Australia
Sheep farming – Australia
Farm life – Australia

636.3145092

Cover photograph of farmer by Steve Bowman, courtesy of Fotolia
Cover design by Blue Cork
Map on p. ix by Alicia Freile
Internal design by Midland Typesetters, Australia
Typeset in 12pt Sabon by Midland Typesetters, Australia
Printed in Australia by Griffin Press, an accredited ISO AS/NZS 14001:2004 Environmental Management System printer

Penguin Random House Australia uses papers that are natural, renewable and recyclable products and made from wood grown in sustainable forests. The logging and manufacturing processes are expected to conform to the environmental regulations of the country of origin.

For my wonderful daughters:
Michelle, Helen and Jennifer.
And Heather Hale and Robert Macklin:
generous, inspiring and true blue

CONTENTS

Map		ix
1	THE BOY	1
2	THE PRESSER	15
3	FLYING PIGS, FLYING RAMS AND BUSTED OVENS	39
4	MAD JACK AND THE BIRD OF PASSAGE	51
5	THE GIRL AND THE HONEYEATER	60
6	THE SHEARING CONTRACTOR	68
7	JIMMY G AND THE BALD EAGLE	80
8	BRONCO AND ZULU	101
9	THE JEWEL AND WALLACE	111
10	COUNTRY GIRLS AND A CRAZY COP	131
11	A SHEARER WHO WATCHED EAGLES	140
12	MARILYN	151
13	A BALMAIN BOY AND A FIGHTING SCOT	155

14	HAPPY JACK'S LAST POST	167
15	THE LAUGHING KIWI	193
16	THE GYPSY COOK	208
17	THE MOTHER	223
18	THE CITY AND THE BUSH	229
19	THE BROLGA AND SCREWJACK	236
20	THE TIMES THEY ARE A'CHANGING	251
21	DICK, THE EAGLE AND RITZY	255
22	JOE BLAKES AND A RAMBLING POMMY	266
23	THE LORD OF THE FLIES AND FLYING EMUS	271
24	THE WHITE GODDESS	280
25	SNOWBALL, DICK AND THE EAGLE	285

Acknowledgements	293
Glossary	294
Author's Notes	302

1

THE BOY

1950

'The boy has developed bronchiectasis – a spot on the lung. He might not survive another attack like this last,' Dr Woodhill warned. 'You say he's free of asthma in the west. My advice is that you send him to his father.'

I was in my twelfth year, undersized and skinny and just discharged following another stay in Toowoomba Hospital fighting off chronic asthma and chest infection. My mother, Pat, was struggling to raise and educate seven kids aged from twelve months to twelve years, and to manage financially on the proceeds of the cheques Dad mailed while he was out west shearing. At her wit's end, she put her oldest son on the *Western Mail* train – destination Cunnamulla.

I remember the lonesome whistle of the great engine and the lonesome heart of a small boy with sandwiches his mother had packed along with a tin of jelly beans and some treasured comics. 'Don't you worry, missus,' the kindly middle-aged guard assured my anxious mother. 'I'll keep a good eye on him.' And he did.

I loved my mother – her patience, her understanding, her playful teasing – but as the train pulled out, my thoughts turned from her to my father. I lived in constant fear of him: of failing to live up to his expectations of what a small boy could and should achieve; of punishment for misdemeanors committed through accident or carelessness. 'There's no such word as can't!' and 'Your mother never raised a squib!' were his mantras.

As I nervously clutched my *Rover* and *Wizard* comics, I recalled a six-year-old boy hiding behind the barn to avoid a thrashing for accidentally breaking a plate while wiping the dishes; a small boy finding a track through the cow cane in the half-light of morning searching for milkers, in dread of failure and punishment; an eight-year-old asthmatic being shoved out the front gate to take a second thumping from a bully years older and twice his size. Most vividly I recalled the nightmare of a terrified two-year-old being swung up behind his father onto a rearing stock horse. The horse went pigrooting around the mounting yard while the child howled blue murder, before the gate was opened and the boy was swung, sobbing, from the back of the horse to the safety of his mother's arms.

As I sat back in my seat on the train, sweat began to bead. It was mid-summer and hot as Hades in the second-class sleeper Mum had booked. I reached over and opened the window, only to have coal smoke and eye-stinging grit blast me. Quickly shutting it again, I instead tried to focus on the reassuring rock and rhythm of the carriage; the regular drumming of steel wheels and rails.

THE BOY

The two men in my berth introduced themselves and, as I had been brought up to do, I politely addressed them as mister. 'Call me Dan' and 'Make it Joe, son', they corrected. Dan was a talkative young station hand heading west after a holiday in the big smoke; Joe was a weathered middle-aged drover from Toowoomba, wearing moleskins and riding boots.

When we stopped at stations Joe took me to feed and water his kelpie dogs in the dogbox at the rear of the train. They yelped at his approach and I was allowed to pat them through the bars. Afterwards we went to the dining car together, an adventure for a lad who had rarely eaten anything but his mother's wholesome cooking. I felt embarrassed when Joe shouted my meal, as Mum had given me the money to pay for it, but he insisted.

When we had eaten we made our way back to our sleeping berth. The terrain was flat west of the Great Divide, but the train rarely reached thirty miles per hour and the journey would take about twenty-four hours to cover the 500-mile distance. Stirred by curiosity, passengers walked the corridors of the swaying brick-red carriages, greeting fellow travellers.

We stopped at sidings where railway gangs lived in government houses with wives and kids. The kids and some of the mums ran to the train, joining the fettlers shouting an appeal for reading matter. 'Paper!' they yelled to the passengers, who responded by thrusting the newspapers and magazines they had finished with into grateful hands.

Towns were far apart: Dalby, Chinchilla, Roma, Mitchell, Morven, Charleville, but finally we reached Cunnamulla, where Dad met me off the train with a handshake. Perhaps

I longed for a hug, but an outward display of affection between males was outside Dad's curriculum at that time. Nevertheless, it was the beginning of a new relationship: a tentative step towards a trust which would take years to develop into mutually understanding mateship. For the next couple of years, until I consistently earned a living, I became his 'shiralee' – the swagman's bundle. I was his responsibility.

We travelled by taxi, as most shearers did in 1950, to Dynevor Downs – a vast sheep station belonging to the Kidman Empire. Although the property was rabbit infested, a bountiful wet season had grown a mantle of green grass and herbage over the paddocks and filled the nearby lakes, which hosted seagulls, pelicans, storks, ducks and waterhens. Red kangaroos and wild pigs abounded.

The room Dad selected opened onto a narrow verandah. There were two stretcher beds with a small table between, but no wardrobe, so we hung a few things on the nails and hooks on the walls and left the rest in our ports.

With our room in order I followed Dad diffidently around the quarters, where he introduced me to a few old mates. Jack Muldoon and Tiny Hazelgrove were 'gun' shearers – legends in the shearing fraternity. They usually took the lead, often shearing over 200 sheep in a day. Word of their tallies had come to me through shearers calling in at home in Toowoomba, and now I stood in shy reverence as I shook hands hardened and enlarged by years of hard work. Jack Muldoon towered over me, tall, lean and bald, but Tiny was only a few inches taller than I was. Short in the legs and powerful in a long upper body, he was known as the 'mighty midget' and the 'pocket battleship'.

Next Dad introduced me to Bob Dougherty, who was the Union rep; and Harry, a nineteen-year-old whose athleticism, I would later discover, belied his weakness for drink.

That night some of the men moved their beds onto the verandah or right outdoors to catch any breeze in the stifling heat. I lay in bed in our room with an aching hollow in my child's heart. For the first time I had begun to really care for my father, but if he heard me sobbing into my pillow for my mother and home life, he never said so. That was his way: I had to grow up tough.

Over the next few days I tried to help out in the dusty manure cloud of the sheep yards, but was told too often to 'Get out of the bloody road, boy' or 'Don't play with the pup, boy – she's learning to work. You'll bloody well ruin her.' I had become 'the boy'.

The cook was a cranky old coot who wouldn't accommodate a kid's desire to please, so I stayed out of his way too and invented solitary games. I raced marbles representing race horses down a ramp, and whittled tiny cricket bats and boats, racing the boats in the bore drain and playing cricket matches using a marble as a ball. Heroes like Bradman and Miller would score centuries and send the stumps flying in test matches against the likes of Alec Bedser and Len Hutton; however, playing against Queensland in my miniature version of the Sheffield Shield they were not so potent and Queensland always won.

My mother, who had spent five years as a governess on outback stations, had packed the necessary school books for me, and Dad had placed them neatly on the table in our room. He ordered me to obey Mum's timetable of study while he was shearing, which God knows I tried to do, but the latest *Rover* and *Wizard* that arrived on the mail truck took precedence.

It was only the powerful young wool presser, Jim, who was kindly and understanding of a lad's needs. Stripped to shorts and boots, Jim towelled salty sweat from his shining face and welcomed me to his wool room with a smile and a word, and taught me how to sweep up the locks and weigh and brand bales. The latter was a responsibility I appreciated; it made me feel useful in a grown-up's world. I goggled at the muscles in Jim's nuggety twelve-stone frame, and doubted my skinny five stone could ever grow to such proportion and power – yet ambition was born.

At smoko, I squatted on a wool bale and tucked in to patty cakes and a fruit cake flavoured with cocoa called a brownie. This was washed down with a mug of tea sweetened with condensed milk and sugar. Knowing a small boy's place was to be seen and not heard, I listened keenly to the jokes and yarns, laughter and sardonic complaints that circulated about the wool room, while trying to ignore the four-letter words and blasphemy. I noticed that many of the older men – my father included – rarely swore except for the occasional 'bugger' and 'bloody', which wasn't really considered foul language. The same men – to a man solid unionists and some of them ex-diggers – would pull the foul-mouthed brigade into gear for vulgarly disrespecting women. 'Don't you have any sisters, son?' or, 'I'll warrant you'd get a clip around the ear if you talked like that in front of your mother' were common reprimands.

I've since heard it argued that woolsheds were simply 'factories in the bush', as Henry Lawson described them, but in contrast to the city's rush hour, after the bell rang and the blades ceased their chatter, shearers often relieved their backache with an after-dinner walk, enjoying a yarn and a smoke as they strode in step with nature beneath the immense and silent arch of a starlit outback sky.

Gradually, over five weeks at Dynevor Downs, I became accepted as part of the landscape: walking and listening, offsiding for Jim, learning card games and tricks, gaining confidence and security as I tagged along pig hunting on a weekend. By now I was completely unaffected by the asthma that had plagued me. It had simply vanished.

When the team cut-out at Dynevor we went to The Lakes for a month. The team was astonished by the vast expanse of water, with waves lapping and flocks of seagulls 500 miles from the sea. Commandeering a rowboat, the workers rowed for miles on weekends, diving overboard and splashing and playing, like kids holidaying at the seaside. On two weekends Dad borrowed a station truck and we drove to the opal fields, deserted long ago when the price of gems collapsed. Shearers and rouseabouts fossicked through the mullock heaps and I recovered discarded potch and colours, collecting them in a tobacco tin as a gift for my mother.

A few more sheds brought the June break at the end of the financial year, but the first week in July found my father and me again beginning a run out west. By now the 'shiralee' had become handy, if not paid. I could 'pick-up' fleeces and throw them neatly on the wool roller's table to have the rough skirt removed; I could skirt fleeces passably, and was taught to further grade the skirtings. 'C'mere, boy!' a rouseabout would command. 'Show yer form while I go for a crap.'

Weeks passed quickly: sign on, cut-out and travel – sometimes hundreds of miles – to the next shed. And then it was the second Tuesday in November when the team downed tools to listen to the Melbourne Cup. I was happy that the run was drawing to a finish, and even happier when I drew Comic Court in the shed sweep and won five quid. My whole family would meet the *Western Mail* in Toowoomba, and I would have money for Christmas gifts.

When my mother enfolded me in her arms I was so robust that she placed her faith in prayer and again enrolled me at the Christian Brothers College. Scholastically I did well, but the old enemy would not be denied, and three months later an ambulance sped me to Toowoomba Hospital with asthma and viral pneumonia.

In the hope that Brisbane's climate would better suit my health, after discharge I went to stay with my loving widowed grandmother and bachelor uncles, Kevin and Gerald. Like many working-class men, their educational opportunities had been truncated by the Great Depression, but they read and were gifted and humorous storytellers – warm and wonderful uncles indeed! I attended Marist Brothers College for a few months before nagging allergies developed into an asthmatic assault which sentenced me to weeks in the Royal Brisbane Hospital.

Before long my father and I went west again for another shearing run. Then Dad, endeavouring to unite the family permanently, bought a dairy farm near Rosewood, west of Brisbane. Father and son milked thirty cows twice a day by hand, separated the milk, fed the pigs and calves, and rounded up the multiplying farm chores. It was such a big work load that Dad decided to modernise and dived deeper into debt by installing milking machines and buying a new Ferguson tractor.

Captaining the Fergy, I was as proud as Punch as I ploughed our paddocks and contract-ploughed for neighbours. However, asthma again struck and Ipswich hospital claimed me for a couple of spells. Lying in bed for weeks I felt guilty for not pulling my weight on the farm.

The farm went bust, and Dad put a deposit on a large Queenslander at Wilston, an old Brisbane suburb. For Mum it was a dream realised: she had grown to love the city as a girl when her father transferred around Brisbane's hinterland and suburbs in his job as a stationmaster. She had lived seven years at Wellington Point, just south-east of Brisbane, after her father died and her mother, who had been a schoolteacher, was appointed stationmistress there. It was only the Great Depression that had driven twenty-year-old Pat west in search of employment. Now, not only would she be living in her beloved Brisbane, her children would also get a sound Catholic education. And how she loved shopping at TC Beirne's and McDonnell & East in the old city!

By contrast Dad, who had been raised on the family sheep and wheat farm near Moree, detested city life. Although he was educated at a boarding school in Glen Innes until he left at sixteen, he had never accepted the value of broader learning and tried to mould me in his own image, believing that a wide knowledge of sheep and cattle, horses and working dogs would 'see a man through'.

I was fourteen when Dad and I again boarded the *Western Mail*. We worked for the United Graziers' Shearing Cooperative around Cunnamulla, and then signed on with shearing contractor Bob Pickersgill in the St George district. I was a rouseabout (known as a rousie), and Dad once more found himself forced by financial circumstance to earn a quid through back-breaking hard yakka – far from the family he loved and lived for. He had softened by this stage, especially with his younger four kids. Dad nicknamed his last born 'Chirp', for as a toddler Kevin had loved to sit cockily on Dad's shoulders as he hurried around doing the endless dairy chores – a bonding removed beyond measure from the terrifying double on a bucking horse I'd had.

Over this period most of my earnings contributed to the family income. The Union ensured that I was paid adult wages as a picker-up as the men in Queensland sheds reckoned anyone doing a man's work should receive a man's pay – an opinion they enforced. With the small amount of money I retained for myself I bought boys' books of adventure, and moved on to lively accounts of popular Australian history with Frank Clune and Ion Idriess. I went adventuring in Africa and South America with Rider Haggard and Colonel Fawcett as my guides – little imagining that in my own country remained wide tribal lands waiting to be explored, inhabited by Indigenous tribes that had yet to come into contact with European civilisation.

The job was repetitive and before long I became an automaton. Every now and then three or four fleeces would come off together and a bossy shearer would bellow, 'Wool away, boy. Wake up!' How could the sweating shearers know that the boy gathering and throwing fleeces was but an avatar for a young man who was opening the batting for Australia, boxing for the world middleweight championship, or fighting off savages in a lost city in the Mato Grosso?

Dad and I parted for a while when he went sheep dealing with his lifelong mate Spanner Hayes, an ex-drover, boxer, soldier and poetry-quoting adventurer who had made his money contracting New Guinea natives for work in the plantations. I signed on as a rousie with a St George shearing contractor, Charlie Stewart, an unlikely character for his calling. Charlie's craggy features and lanky form clad in khaki overalls brought Abe Lincoln to mind. Like the great president, Charlie was a deep reader who loved to quote the classics; he was philosophic, good-humoured and generous. Inevitably he was known as 'Abe', or 'The Long Fella'.

When Dad got back we signed on for ten shearers at Thomby station, about twenty-five miles from St George. A couple of weeks after the cut-out Dad and I joined several hundred enthusiastic spectators at the annual Thomby Rodeo. I tried to ride a bullock for ten seconds for five quid. In four seconds the beast bounced me off the rails and spread-eagled me in the dirt. Dust-covered and dazed I wondered why the crowd was applauding me, until I realised they were cheering the bullock as he performed a victory circuit, bucking, tail-swishing and bellowing – and heading my way. Helter-skeltering over the rails, I drew more jeers than cheers. Dad checked that I was physically alright, and then summed up: 'You forgot to watch his head, lad; always follow the head and keep your balance.'

That night Spanner Hayes offered more advice. 'If rodeo's not your long suit perhaps boxing might be your style.' With no kids of his own, Spanner had taken me when I was eight to the Brisbane stadium to see Alf Wells – a thirty-eight-year-old ex-POW – fight young Jack Daniels to a standstill over twelve rounds. I was immediately hooked. Now, seven years later, he decided it was time to teach me a thing or two about the sport I loved.

I took to boxing straight away, and though I was small for my age I was strengthened by work and decided to enter some local boxing tournaments. With Spanner in my corner I managed to box the ears off two school kids who were my weight but a couple of years younger. Next I tangled with 'Tibby' Gibson, who was my age and also a picker-up in a shearing team. He looked pretty big to me.

My long reach, left and right shuffle steps and jabs, and the right crosses that Spanner had taught me failed to keep Tibby at bay. He came in hell-bent, flailing like a windmill in a whirlwind. At the first taste of blood I forgot Spanner's advice to 'always keep cool and watch your opponent's feet' and I went at him like a threshing machine. When the ref pulled us apart and called it a draw at the end of three rounds, the crowd went wild and threw a shower of coins inside the ring. There was even a fluttering of ten-bob notes and a few flying quids. There was money in the game!

The next morning Tibby and I ran into each other up the street. We were both wearing shiners and fat lips. Tibby said if he hadn't hurt his thumb in the first round he would have knocked my block off. I said he couldn't knock my sister's hat off. We circled, neither of us eager to throw the first punch, but neither willing to take water. Fortunately, Poddy Waters, an Indigenous local footballer and workmate of mine, happened along. 'Settle down, boys,' he said. 'Last night you split thirty-odd quid between you, and now yer want to flog each other for sweet Fanny Adams. Wake up to yerselves! Shake hands now, and save the return bout for the next tournament when yer can get a quid.'

Poddy was making sense, and he knew we were both hoping for an honourable way out of more punishing conflict. We shook hands. We never did have a return bout.

Over the next three years 'the boy' worked about south-west Queensland as a rousie in shearing teams and on sheep stations as a jackaroo and station hand.

In June 1955, Dad bought a Vanguard ute, and we drove to the north-west to Barcaldine, where Dad steered to a flourishing ghost gum by the railway. He proudly introduced me to the 'Tree of Knowledge', the bush worker's symbol of unity since the great shearers' strike of 1891 when the pioneers rallied in its piebald shade to form the Labor Party. Sixty-four years had passed since that watershed in colonial history, but to my youthful mind it seemed like aeons: as far away as Lawson's 'Harry Dale' and Paterson's 'Clancy', or Nellie Melba's triumphs, or the tragedy of Gallipoli. As distant as the rebels at Eureka laying a foundation stone for Australian democracy, or the tragic deaths of young Les Darcy and Phar Lap. Now, more than sixty years later, that journey to the north and all the years between seem but a breath away; and the heroic battlers of ninety-one breathe companionship.

After working around the Longreach and Winton districts for several months we were employed at Beryl station late in the year, when word came over ABC radio that negotiations between the Australian Workers' Union and the United Graziers' Association had collapsed. The situation was similar to 1891: wool prices were falling and graziers wanted the so-called 'prosperity lading' deleted from the Award. Shearers wouldn't wear it. Teams were walking off, and the big strike was on. I was as happy as Larry: I'd soon embrace family life with Mum and the brothers and sisters I hadn't seen for six months.

2

THE PRESSER

The strike of 1956, which was supported by the Australian Workers' Union, soon became as bitter as the Shearers' War of 1891. The Shearers' War had been an epic struggle that had cemented the labour movement's determination to bring social justice to Australia through parliamentary representation, and make the young nation a society proudly based on a 'fair go' – at least for white-skinned male inhabitants. The 1956 strike, by contrast, was far less visionary. It whipped the wool industry and its financial dependencies through a long, expensive struggle, and fed the newspapers a feast of sensational headlines of outback confrontation, threats and brawls from January to October.

Dad advised me to 'stick with the Union but keep out of the firing line'. Accordingly, I jogged up and down the peaceful hills of Brisbane suburbia delivering bread for the friendly firm of George Bott and Sons, Bakers, of

Wilston, while industrial conflagration raged west of the Great Divide.

On Saturday mornings I attended Australian Workers' Union meetings at the Union's headquarters, where I heard rowdy echoes of the social and financial strife consuming the western merino country. The terms 'new-raters', 'old-raters' and 'fifty-sixers' entered the outback lexicon, while 'scab' and 'blackleg' (strike breakers) were resurrected. The transport unions refused to handle 'blackened wool', which was wool harvested by blacklegs. Unionists shore for the old rate of Award wage, while fifty-sixers sweated for the lesser new rate. Most of the outback towns swiftly developed a scab pub, patronised by the new-rate 'black' shearers. Such waterholes prospered during the strike, but the stench lingered, and for years afterwards Union men and supporters wouldn't push through the bat-wing doors of black pubs.

Most of the new-rate clip was harvested by teams of 'tomahawking' learners (so called because their work was so rough and bloody it might have been done with a tomahawk) and by Kiwi imports and scabby unionists employed by the graziers' co-operative shearing companies. A handful of thugs were imported to stand over the Union men, but they were matched more often than not by the bush knuckle-men in some memorable all-in brawls and one-on-one battles.

Union members who scabbed were branded as 'germs', but the graziers' sons who temporarily took up the blades to help out were seen as legitimate soldiers of the other side. Even lads who swallowed the 'learn to shear and get rich quick' propaganda ads were viewed as men who might be educated to Union principles once the strike was over. Likewise the imported Kiwi strike-breakers.

An essential condition of strike settlement – agreed upon by the United Graziers' Association and the Australian Workers' Union (AWU) – was that all new-raters be 'whitened'. This meant that they took AWU tickets before sign-on; it was follow Union rules or roll your swag. The United Graziers' Association needed professionals – and learners who showed promise – to remain, as did the AWU, which was always endeavouring to increase membership. A combination of good seasons and the wool boom had encouraged graziers to breed record sheep numbers through the 1950s, and there weren't enough shearers to keep up with the work as it took years of practice to develop the skills of a professional shearer and required innate hardiness and determination.

Wool was the nation's major export, so it was in the national interest to return the wool harvest to full capacity. Graziers needed to pay their mortgages, while storekeepers, who had given credit to the local strikers, wanted unionists back at work to square family tucker bills. The wiser heads in the shearing fraternity saw the advantage of peaceful solidarity in the industry and they set about educating the fifty-sixers, who had been courageous enough to weather the storm, in Union rules and principles.

The major contractors, UNGRA and GRAZCOS, usually placed a phalanx of solid Union men in each team. These were seasoned shearers who could be relied on to cool the hot heads (for a few hard-cases – usually single men – were on the lookout to settle old scores violently).

A scab list, falsely claiming to be issued by the AWU, even went the rounds, causing ill-feeling and sometimes rows and punch-ups. The list was inaccurate, accusing solid men – along with the guilty – of scabbing. The wiser heads knew it was the product of perennial troublemakers and burnt copies as they came to hand.

The strike might be over officially but malicious 'pointers' continued to spread fact and fiction and innuendo. While peace was usually maintained till cut-out, some crackerjack fights broke out once a pub was reached. On one notable occasion animosity was chained only till the boundary gate was shut; then two car loads erupted and ten shirtless shearers punched each other back and forth across the Mitchell grass plain until they were bloody, black and blue. Exhausted, and seized by the terrible thirst that a twenty-minute stoush in the outback summer sun induces, they called a truce and drove forty miles across the plains to the Gilliat pub. There they guzzled a few pots before surging out the back to knuckle on.

The publican was an ageing, raw-boned legend known as the 'Downs Tiger', named after the venomous snake. The Tiger would rather have a fight than a feed. Roaring into the fray he knocked shearers skew-whiff left and right. 'Now get back inside,' he bellowed. 'Youse come here to drink, not to fight. I've been waiting three weeks to get a quid out of you bastards. Youse all owe me for grog I've sent out on the mail truck. Now pay up and get a gutful of piss – and then youse can fight till the cows come home.'

With the strike settled in a rare points win to the Unionists, I returned to the south-west in April 1957 and became a wool roller, followed by piece-picker (a rousie who sorts the lower grades of wool, removed by the wool rollers) in Richie Jack's team near Bollon, south-west Queensland.

Richie was known as Sack 'em Jack, a nickname he had earned during the 1930s. For me it became hard to associate the stammering, affable grey-haired man with the ruthless overseer who had earned his reputation during the grim years of the Great Depression. Old hands, however, still recalled bitterly Sack 'em Jack firing shearers who were struggling family bread winners.

Younger blokes who hadn't experienced his harshness firsthand would chuckle as they passed on a tale telling how Richie had gained another nickname: the taxi-loader. Apparently, after firing four men in quick succession, one of the men told him, 'I'm going to the homestead to use the phone to contact the Union organiser. He'll pull you into gear quick smart.'

'No need to phone, s-sport,' Sack 'em replied, 'you can s-s-see him when you get to t-town. Get your cheque. You'll make up a taxi load.'

Of the eight men on the board half were learners – class of fifty-six – which meant the team was shearing only as much wool as an average five-stander. Even so the wool presser, an overweight fifty-sixer, soon threw in the towel.

'Heart the s-s-size of a p-pea,' Sack 'em stuttered as he handed the fat man his cheque. He offered the man-sized job to 'the boy'. Turning to me he said, 'You're the p-presser, young Blunt! Want it?' Did I ever!

Being a wool presser involved putting the wool into a wool press to make bales, and it was the heaviest work in the shed. Under the Pastoral Award pressers could work unlimited hours except on weekends. However, it was a matter of pride among professional lever-men to keep the wool away within the bell hours. Professional I was not. I was eighteen, fit and hardy but underweight and under-powered for the job. Muscle-sore and dog-tired I plodded to the shed each night after tea with a carbide light and a proud can-and-will determination. In the gloomy interior I loaded and tramped wool and swung on the lever for a couple of hours to lower the wall of fleeces prior to the next morning's bell.

As I passed the boss's hut, Richie Jack, who was relaxing on the verandah, removed his pipe, blew a cloud of pungent Erinmore aroma, chuckled, and called to me a stammering, 'A scratch team got you b-bogged? Never thought I'd see you doing a F-F-Florence Nightingale for eight snaggers, Chapman.' Bill Chapman was the doyen of wool pressers – a peerless giant of a man – and pressers who had to work at night to catch up were jokingly referred to as Florence Nightingale (aka the lady with the lamp). Thereon Richie Jack dubbed me 'Chapman', or AJ (my initials); others called me 'the presser'. 'The boy' was forever gone.

In April 1958, after gaining experience as a wool roller, drover and roo-shooter for twelve months around south-west Queensland, I met Peter Hargreaves at a shed out of Goondiwindi. Peter was a young solo-slaughterman-cum-shearer from Wangaratta and although he was shorter than me he was a stone and a half heavier and strong as a scrub bull. He had done a bit of boxing and volunteered to be my sparring partner to prepare me for an upcoming tournament in Goondiwindi.

Each night a few of us would take carbide lights to the woolshed after tea, where I'd box rings around a dogged, pursuing opponent. Although I pulled my punches Peter ended each session with claret dripping from his nose and reddened ribs and face. 'I'll catch you one night,' he'd say, grinning while we shook hands. Our final spar ended when he caught me fancy dancing in front of the open wool press door with a shoulder charge that knocked me inside the box. Trapped, I tried to fight back while Peter thumped me vengefully till I dropped, winded and bleeding. Stepping back, he chuckled and pulled the gloves off. 'That's our last spar, Presser. Now we're square.'

The bout in Goondiwindi fell through when my opponent didn't turn up. Instead I boxed an 'exhibition' with a lad from Roma whose opponent also didn't show. We must have put on a good performance, for as well as bloody noses we drew a shower of coins and a few notes.

Remaining good mates, Peter and I lined up work with UNGRA Shearing Co-op in Hughenden, while Dad took on overseeing shearing sheds for a Goondiwindi contractor. In June we drove up the coast in Peter's Holden ute. We passed a week lolling through the sleepy coastal sugar towns. 'The further north we go,' Peter observed, 'the girls get prettier and everything moves slower. There's more push bikes than you can poke a stick at – and I'll tell you, all this pedalling sure gives a girl good legs.'

In Townsville we went to the Olympic pool to watch the Empire Games team in training. It was an easygoing era: swimming champs were amateurs and Aussie heroes, not millionaire celebrities barricaded by millionaire agents. My Box Brownie snapped friendly shots of Dawn Fraser,

Lorraine Crapp and Gary Winram, before we headed for Hughenden – and a meeting with Big Bob Teitzel, United Grazier's Shearing Co-op Manager for the north-west. 'Big Bob' had been dubbed variously the 'Provo' (slang for military police) and 'Gestapo Bob' by the hardy independent ex-diggers who shore in his teams. A part of Big Bob's martinet policy was to split mates by putting them in different teams. His theory was that if one pulled out or got the sack his cobber would go with him, and the team would be short two men instead of one. However, this didn't apply to everyone, for Bob knew from experience that some old mates and brothers worked together as a condition of employment, and that they were usually reliable, above-average shearers. Thus I was sent to Vuna station, south of Hughenden, while Peter was placed near Julia Creek, 400 miles west of Townsville.

Following Vuna cut-out I went on to Malboona. I was travelling light – my only luxuries were half a dozen books, a dismantled twenty-two rifle and a set of boxing gloves, crammed into a port and swag.

Since the 1890s, when Henry Lawson wrote his vivid descriptions of big sheds and big men about Bourke and along the Darling River, the shearing fraternity had been noted for its individuals and characters. Seventy years on, the Malboona team of thirty men could boast an above-average quota of the same. 'Darkie' was one of those characters. He was an easygoing bloke everyone in the brotherhood seemed to know, like and respect. He was an unobtrusive peacemaker who could defuse explosive situations with a few words, or liven a dull evening with puckish wit and amusing yarns.

Darkie and a few others were having a drink in the Grand Hotel waiting for a taxi to take them to Malboona when two unknown newcomers from New South Wales breasted the bar. After the usual 'G'days' had been extended, Darkie made a friendly inquiry: 'You blokes goin' to Malboona?'

'We were,' Johnny said pointedly, 'until Adolf Hitler's bloated uncle laid down the law about grog in the shed.' He was referring to Big Bob Teitzel's proclamation that anyone taking alcohol onto a station would be dismissed, according to a rule written into the Shearer's Agreement but usually ignored.

'Strewth!' Darkie exclaimed. 'He don't have a search warrant, yer know. If it suits yer I'll go and have a yarn to him.'

'It's too bloody late now, cobber,' Johnny said. 'I told the red-faced Gestapo prick he should have been hung in Munich with the rest of his Nazi butchers.'

'I see what you mean!' Darkie said to close the subject, but Johnny's mate George wasn't satisfied. A stocky, surly-looking character of dark complexion, he stepped back from the bar, hooked his thumbs in his belt and glared belligerently. 'That so! Well, I *don't* see what you mean. Anyway, we wouldn't want to work with bastards weak enough to sign on under Gestapo rules. We won't work with scabs, either.'

At fifty-odd Darkie was twice the challenger's age, and grey and stringy from a lifetime of hard work. Yet, confronting George, who looked rough enough to wrestle a grizzly bear, he was a picture of relaxation. With a fag dangling from the corner of his mouth, he drawled, 'Hang on, mate, and I'll show yer something.'

Champing at the bit for a fight, but puzzled, George stepped close and waited while Darkie dug a World War II overseas service medal and ribbon from a scarred and shiny wallet. Darkie's voice firmed with pride. 'No Gestapo rules here, young fella! I fought the Nazis. I fought the Germans, too – and worked with 'em, before and after the war. Solid men, most of 'em.' Extending the medal he said affably, 'Your old man might have one of these.' He laughed and added, 'Big Bob's trouble is he thinks he's still a provo.'

Johnny intervened. 'You're right, old-timer. Two of my uncles have got those medals, and a few more besides. We'll buy these blokes a drink, George. No hard feelings.'

I had often wondered why ex-diggers, peaceful men and boys whose courage had been tested in the crucible of combat, wouldn't or couldn't talk of their experiences. Later, having become mates with Darkie, I said, 'I didn't know you had that medal, Darkie.' Sounding slightly embarrassed, he said dismissively, 'There are plenty of them about, son. You just had to be in the wrong place at the wrong time; and have enough luck and good mates to see yer through.'

Another larger-than-life figure at Malboona was Snowy Hales. Yarnsters and mythologisers had spread word of his deeds and capabilities about the west well before he shook hands with me at Malboona. He was a gun shearer but his work was none too neat, relying on consistent raw power rather than the fine touch and rhythmic stamina that were the usual hallmarks of a 'dreadnought' (a man who could manage the rare feat of shearing 300 sheep in a day). When he wasn't on the shearing board Snowy wore high-heeled riding boots, perhaps to top up his height but more likely because they reminded him of the carefree years he had spent as a young ringer (mounted station hand) and horse-breaker – jobs he still preferred at times. He was said to be the strongest man west of the Great Divide, a reputation he maintained by exercising regularly with weights, calisthenics and running.

Snowy's transport was a Holden ute. At some sheds he erected his tent a little distance from the quarters, preferring the solitude of wind and sky and birdsong to the close company of the huts. A quiet loner with a boyish, reassuring smile, Snowy always emanated a powerful presence – an aura similar to that of a dozing blue heeler cattle dog on guard on the back of a station ute. Occasionally he would break out on the booze and go on the tear, when it was said that he emptied public bars quicker than a striking taipan. It would take three or four brave and burly coppers to yard him.

Snowy's wife had died a few years earlier and he shore chiefly about the south-west to keep in touch with his two dominant passions: the student daughter he boarded at the Charleville convent, and opal gouging. Come weekends, be it chilly winter or blazing summer, he could be seen sitting outside his tent burnishing opals in the bright sun, clad only in shorts and riding boots. On occasion he shyly produced photos of his daughter to show proudly to the few folk he allowed entry to the outer perimeter of his personality.

Less than a decade later word went around the bush telegraph that Snowy had been shot by his own hand. Some said that he'd fallen victim to the 'Spanish dancer' (cancer); others that a bad bout of the blues, following a prolonged breakout on the demon rum, had pulled the trigger. Doubtless there were other factors – and many combined on the fatal day.

Malboona could boast only two shearers' vehicles beside Snowy's. These utes belonged to coves out to save a quid: shearers reluctant to go to town despite regular Friday night and weekend requests from eager workmates. Thus for most of the four-and-a-half weeks of the contract the team was confined to barracks and, when not working, had to battle boredom off their own bat. Poker was the bushman's favourite, but other card games like five hundred, whist and cooncan also helped to fill the hours.

I played chess with big Karl, my German mate on the Ferrier wool press, and with David, a gangly young learner shearer whose parents were dairy farmers from south of the border. He'd had a private school secondary education and was saving to study applied science at university – a resume which might have made him an outsider but for his sharp wit and reserved camaraderie. Snowy took the youngster under his wing, gave instruction and shore the difficult sheep while still nearly doubling his protege's tallies. After David shore his first hundred, Darkie observed wryly, 'I don't know what yer want to be a scientist for when yer can earn a quid slaving over hot bodies with yer arse above yer head shearing maggoty sheep!'

After tea on weekends blokes gathered in the warmth and flickering light of the traditional campfire. Horse races and shearing were the main topics, while debates on cricket, boxing, footy and women were prominent. Prime minister 'Pig-iron Bob' Menzies, 'Black Jack' McEwen and their Tory cronies were roasted, as was 'warmonger'

Churchill, whose hand at Gallipoli was forever branded on the Australian psyche.

'Old' Snowy (dubbed old to distinguish him from Snowy Hales and Tall Snowy) often held sway with stories of drama and humour, but when it came to stretching the truth Darkie had no peer in a shed that boasted half a dozen shameless liars. 'Now, Wingy Johnson,' he'd begin. 'He was a one-armed little blighter with an evil face; the dirtiest and worst cook on the Warrego, and so greasy dogs licked his shadow. It goes without sayin' that he woulda been sacked on the first day except for his famous rissoles.' At this Darkie rolled his eyes and smacked his lips. 'Talk about be-yootiful. I can taste them rissoles yet! Blokes boasted about 'em for years – but Wingy would never disclose his recipe. He'd serve 'em up for breakfast and again for tea – and the boys would tuck in to the same burnt mutton and stewed veggies and blowflies in the soup every day and never think of complaining – as long as they knew them rissoles was coming. And then one afternoon a rousie runs to the huts to get a shearer's pipe. He looks through the kitchen window, and cops Wingy mass-producing rissoles. A handful of mince shoved under Wingy's stumpy armpit, a roll of the shoulder and into the pan went the tastiest rissoles ever made. It was a sad day: Wingy had to go, and his recipe went with him.'

The newcomers not yet awake to Darkie's phenomenal powers of exaggeration listened in wonderment, while the initiated chuckled and said, 'Tell us another, Darkie.'

Darkie didn't crack a smile as he took his time rolling a smoke. 'Wingy's brother, Harold,' he started, 'was known as Poisoner. Judgin' by his name yer might think Poisoner was a terrible and dangerous cook; in fact he was the best babbler on the Barcoo [cooks were usually referred to as 'babblers', an abbreviation of the rhyming slang 'babbling brook']. He got his handle because his hobby was keeping pet hoop snakes in a box in the kitchen. As all educated people know, a hoop snake puts his tail in his mouth and rolls after his victims. In fact, hoop snakes are the reason I gave up droving: a big hoop snake can outpace a galloping horse, spring off the ground, lasso a drover and have him fanged and dead in seconds. I happened to be there the day the Poisoner kicked the bucket.' Fixing his gaze on the young rousies, Darkie declaimed, 'Poisoner laid a lotta blokes in their graves with hoop snake venom – blokes who got offside with him, which was easy to do. But he was a babbler right out o' the box, and most of his victims were rouseabouts, so they weren't much loss.'

Lighting a cigarette and taking a drag, Darkie continued, 'As yer know, anyone can tell if a bloke has been killed by hoop snake venom: he kicks off his boots and puts his big toes in his mouth and rolls fifty yards or so before he keels over, as stiff as steel and bright blue. We were heading to the mess for dinner when the Poisoner comes rolling towards us like a runaway wagon wheel, and flops over, stone dead, stiff as steel and bright blue. We knew right away the Poisoner had accidentally been bit by a hoop snake. We couldn't straighten him out, so we got bars and shovels and buried him in a circular grave right there. For years afterwards, as a mark of respect, rouseabouts in passing would piss on his grave.'

When we weren't sharing tall tales, the campfire philosophers would take a turn holding court. Perhaps sitting under an open sky watching the hypnotic heart of a campfire inspired spiritual contemplation; at any rate it wasn't unusual for discussion to turn to the meaning of life and God's role in creation. David declared that chaos ruled our lives because God was dead; a surprise to me, albeit the influence of my Catholic upbringing was well on the wane. Karl explained his belief in Duality: that God and Lucifer were opposing expressions of the same deity, locked forever in eternal struggle, and the good and evil in mankind's soul was an inevitable reflection. Only later, when we had become close, did Karl confide to me that as a boy he had served a term in hell on earth as bombs and shells blew German towns and cities apart; of the stigma and shame his people had to bear as Hitler's atrocities were revealed to the world; and the submission and abuse he had learnt to endure through the Russian invasion and occupation.

With a full team in camp, Malboona was a lively place on the weekend. Saturday mornings we caught up on washing and personal chores. Frankie took SP bets and, after dinner that first Saturday, loaded ten or twelve starters on to his ute and headed to the Corfield pub forty miles away. They returned in time for tea, half full of beer and full of cheer, with a week's supply of booze on board. Sack 'em Jack Butler, the shed overseer (also dubbed 'Sour Jack', as distinct from the original Sack 'em Richie Jack), noticed the booze, but was wise enough not to mention it as half a good team would be impossible to replace if they pulled out.

After tea Darkie laid out a tarp and mustered four or five carbide lights, while Old Snowy produced the kip and pennies for a two-up game. The night was nippy and punters warmed themselves around the blazing fire. Laughter and shouts of 'Two bob on the head' and 'He's tailed 'em!' were interspersed with 'C'mon, back yerself or pass the kip.' Now and then Old Snowy would call:

He's headed 'em twice but he can't do it thrice –
Take my word, boys, I'm a scholar.
If you haven't been set, I'll cover your bet,
I'm backing a tail for a dollar.

At seventy-two, Snowy Robbins had seen life incrementally transformed by the railway, telephone, motor vehicle, silent movie and electric light. Now, Australians bought radios, gramophones and home appliances, went to the talkies and holidayed on the coast. Malboona and other big sheds were the tail-wag of a vanishing culture where isolated people had to find their own solutions to boredom.

Old Snowy was one of those rare blokes who had entered life with the blessing of a joyous heart and glowing spirit, and a mission to raise morale in others. He had survived and prospered through good times and hard, including two world wars and the Great Depression. Lean and straight with wispy white hair, the retired dreadnought was a long-time legend. He was a Union man who proudly earned his living as a piece-picker, and freely exercised his right to criticise the Union and any other organised body. The old maestro was still capable of wrapping a rheumy hand around a shearer's hand-piece (known as a bog-eye), and stylishly removing the fleece from a woolly to show a learner how easily it could be done.

David watched him closely, learning all the while. He confided to me, 'It's lovely to see the way he does it: poetry in motion; he massages the wool away. And he's a lovely old man, too – always ready to advise, and always smiling.'

Years later it struck me that Old Snowy had consciously fallen into the role of a kindly mentor, a responsible survivor unobtrusively passing on socialist philosophy, principles and practice through example, story and song. On a few occasions the old entertainer pulled out an ancient mouth organ and a pair of soup spoons and rendered old familiar tunes such as 'Two Little Girls in Blue' and its sequel 'After the Ball' in a cracked but lyrical light tenor.

Old Snowy would get the boys in chorus with war-time ballads: 'Lili Marlene', 'Knees-up Mother Brown', and the 'White Cliffs of Dover' – before whacking out a few songs that projected some of the real feelings of the front-line womanless diggers:

I love you in your negligee,
I love you in your nightie,
But when moonlight flits across your tits –
By God all bloody mighty!

The lean and hungry days of the Great Depression still troubled the memories of many Aussies, and Old Snowy with his rebel heart cried out for social justice through his creaky, tuneful rendition of Buddy Williams' popular historical song 'Wingie the Railway Cop' and Tex Morton's banned 'Sergeant Small'. He got a lively response with hand-clapping and calls of 'You beauty, Snow! Bore it up the bastards!'

When he moved on to Banjo Paterson's iconic ballad, with a few boozy voices harmonising to his mouth organ,

Karl queried, 'What is this "Waltzing Matilda"? It is a good tune, but bloody sense it does not make.'

'Yeah, you're right, Karl,' I agreed. 'It mightn't make much sense but it beats the hell out of tipping your cap and singing "God Save the Queen".'

On Sundays fighting men and runners still occasionally challenged for a wager. I pulled my gloves out, and most of the boys trooped to the wool room on Sunday mornings to encourage with cheers and boos. Tall Snowy claimed his mate, Dinny Carew, a young rouseabout, would have won the Australian featherweight title as a teenager if booze and a dislike of training hadn't knocked him out. I called, 'Bluey, how about a light spar, mate?' to a laughing freckle-faced kid whose skin was almost as ruddy as his hair. The eighteen-year-old picker-up said he had never had a fight in his life but 'I'll give it a go – if you don't punch hard.'

'I'll be as gentle as your mother,' I replied, believing I could carry Bluey with one hand behind my back and display some fancy footwork, while treating the spectators to a lot of laughs.

'You've never met my mum,' Bluey quipped. He was half a stone lighter than me and nowhere near as fit or as strongly muscled, but he whipped into the fray like a southpaw whirlwind, weaved inside like a professional, and hammered me into a circling retreat. I tried to keep grinning as I ducked and dodged and gave ground.

Giving voice to instinct, the twenty-odd onlookers rattled the old iron roof. 'C'mon, Bluey, bore it up 'im! Punch the bugger clean out of his soul-case.' After a flurry of leather bloodied my nose came cries of 'Cop that young 'Arry!' and 'You little ripper, Blue!'

THE PRESSER

Before the third and final round I concluded that Bluey's carelessness with the truth concerning his ring experience freed me from my promise. I went after my man, putting real heft into my punches with hip and shoulder, but with little effect. Puffing like a loaded steam-train engine, Bluey slipped and rolled the weight out of the blows and buzzed in to attack like an angry red wasp.

Officiating as ref, Snowy Hales called on the crowd to give a decision. A thundering cheer mingled with profane exclamations of admiration left no doubt that Bluey was the punter's choice.

I grinned through bloodied lips as we shook hands. 'Too good, mate. Let's do 'er again next Sunday.'

'Yeah, any old time,' the rousie puffed. 'But only two rounds – or I'll have to quit the fags.'

'Anyway, ridgy-didge, where did you learn to box like Jimmy Carruthers?'

'Me? Like I said, Presser – I've never had a fight in me life! I saw you jokers swingin' and gigglin' like girls in a pillow fight. I thought my sister could beat these pansies, so I says to meself, I'll give it a go. I'm a bloody natural, ain't I?'

'Natural bullshit artist!' I said, laughing. Later, loosened by a few beers, Bluey confessed he'd been runner-up in the New South Wales schoolboy titles. 'The trainer at the Police Boy's Club said I might have a future as a pro if I gave up the booze and fags, so I give up boxing.'

As we strolled back to the huts, chatting, chiacking and joking, Karl said seriously, 'What is with you blokes? I think you the presser's mate – like me – and then to kill him you want, Bluey.'

Most of the boys looked at him quizzically, wondering where he was coming from. Darkie grinned and shifted his drooping smoke to the opposite side of his lips. 'Easy one, Karl. You'll catch on. The newspapers like to write us up as the "classless society". Maybe we ain't got toffs like the mother country, but after you've been around the traps for a few years you'll know that's bullshit. There's top dogs and underdogs here – like anywhere else – and most of us is for the underdog, because we are underdogs. There are dogs piled on dogs all the way down until you get to bottom dogs – they're called rousies – like young carrot top here.' He ruffled Bluey's crop of rusty curls. There was laughter all round, and Bluey shaped up and said, 'Keep an old man's place or you'll cop the same as the presser copped!'

Darkie drawled on, 'We're like dogs because we have to lick arse now and then whether we like it or not. The fact is we've all got to learn to eat shit from time to time, it's part of growing up – and it's better than a biff behind the ear with a rifle butt like you'd cop off the Nazis or Commos. It's getting to like the taste of shit that's bad for yer character. And the good thing about Australia is you don't have to get to like the taste of shit; you can tell the boss where to put his bloody job any time you like – whether yer can afford to or not.

'Now, take a gander at the presser here. He likes to dance around and big-note himself while some mug takes a swing and a miss and looks silly, so when a red-headed kid surprises us all and cuts him down to size we're on his side. But it don't mean the presser ain't still our mate – just because we sling shit at him.'

THE PRESSER

The big German looked more perplexed; and my grin broke into a chuckle. 'That's right, Darkie. Your elucidation has made it as clear as mud for the old mate.'

'There he goes again,' Darkie said. 'Coughing up the dictionary and getting in the last word.'

Clearly I was picking up a reputation for big-noting by dropping in words I had learnt from poems and short stories I loved. I had with me volumes of Blake and Keats, Rudyard Kipling, HG Wells and Katherine Mansfield, and a volume of Russian short stories. The jibes were all in fun: Karl would give me a brotherly clap on the back which nearly floored me, and comment, 'You full of the bullshit. Politician you should be!'

Saturday and Sunday morning smokos were later because breakfast was postponed till eight o'clock so that the cook could enjoy the luxury of a sleep-in – till 5.30am. Snowy Hales and a few of the old hands had the coppers smoking and their washing on the boil before sun-up. After breakfast had settled and the washing was hung to dry, Snowy kick-started a display of outdoor entertainment. He laid out a large railway tarp to show that at thirty-seven years he could still run on his hands and do backflips, tumbles and cartwheels. There were plenty of laughs when some of the young blokes went leg-up trying to duplicate him. Bluey volunteered to stand in front of him and place his hands on Snowy's head, while Snowy crouched and grabbed the rousie by the ankles. 'Stand stiff and straight as a gun barrel and hold yer nerve,' he commanded, before easily raising Bluey's ankles to shoulder height and walking twenty paces with the young man aloft.

The old shearing shed sports of shot-putting with a brick, javelin-throwing with a broom and scratch-pulling were noisily contested before Snowy, swinging a pair of dumb-bells back and forth for momentum, soared thirteen feet in a standing broad jump – half a yard clear of the field. Then about fifteen of the younger blokes lined up for a hundred-yard sprint. Darkie fired a twenty-two rifle in the air, and they galloped over grass and gibbers. Barefooted, Snowy Hales rocketed away, while David, wearing sandshoes and still lengthening his stride, strode past Bluey to finish a closing second. Obviously surprised at his defeat, David eyed his shearing mentor and challenged, 'How about you and I run a hundred yards for ten quid, Snow?'

Snowy smiled and declined. 'I'm not a betting man,' he said. The pressure mounted – already bets were placed. Snowy reluctantly agreed. 'Alright, but let's make it over fifty yards.' They compromised on a match race over seventy-five yards.

David vanished into his room for a few minutes. He emerged bearing a ten-pound note and wearing running spikes – and interest intensified.

'No bet, lad!' Snowy said as seriously as a Jehovah's Witness on a Saturday soul-saving mission. 'Didn't your mother tell you that gambling is a bad habit? Like smoking and drinking it leads to personal degradation.'

David was wrong-footed; his practised aloofness rattled. 'Well, er . . . ah, I don't smoke – and I drink only in moderation.'

This response drew a round of laughter and a gigging from the listeners. Dinny said, 'I'll bet the sheilas reckon you're the life of the party. I'll bet you a bottle of Bulimba beer to a packet of Craven A cigarettes that Snowy shows you his heels.'

THE PRESSER

'You're set – and doubled,' David shot back.

In fact, Snowy Hales was a betting man if he liked the odds, and he usually didn't shy off backing himself in challenges of strength and sprinting. His failure to take a punt on himself indicated to the old hands that he didn't believe he had the young challenger's measure. Snowy's backers had been laying odds; now, suddenly the odds on David dried up. 'Give us the tenner, son,' Snowy Robbins said, 'and I'll see if I can get yer set.'

Waving twenty quid over his head he commanded with the flair of a showground spruiker, 'C'mon, roll up, get on while yer can, fellas. Take a gander at the starters! Snowy Hales here has run 'em off their feet in match races from the Charters Towers to Bourke. He's beat all the guns, and the would-be's and the could-be's, and it's known that only the handicapper beat him in the Stawell Gift.' At this claim Snowy grinned sheepishly and turned away.

'Now turn your attention to the challenger,' Snowy called. 'He's lanky and still growing and so awkward he's got two left feet. Don't let the flash running shoes fool yer! This rooster couldn't run out of sight in a railway tunnel at midnight. He should be ten to one, but I'll take twos, anyone give me twos – six to four?'

After a rally of ribald and sarcastic comments went the rounds, a laughing Frankie Roberts called, 'You're a cagey old bugger, Snowy, but I'll set twenty at even money.'

Observing David's flying finish in the first race, Old Snowy had backed him for a few quid before quietly opining, 'One right out of the box, this fellow. Good hip drive, high knees, straightens up. He's been coached by a pro.' Now he said, 'Twenty quid it is, Frankie! I'll set the other ten.'

It wasn't a cinders track, but the twin wheel tracks made a hard, bouncy surface, the gibbers compressed by decades of iron and rubber-tyre traffic. At Darkie's command of 'On your marks, get set' the contestants dropped as earnestly as Olympic finalists. With the crack of the twenty-two rifle Snowy's short legs shot him off like a tin hare. He looked home-and-hosed at fifty yards, but David had straightened and thrown his head back and his long strides were gaining ground. Grunting, he strode alongside, threw his chest and broke the tape inches ahead of Snowy, while the watchers cheered and yahooed a din equal to five times their number.

3

FLYING PIGS, FLYING RAMS AND BUSTED OVENS

The following Sunday, a dour shearer called Scotty produced a soccer ball and began booting it about. Pretty soon the lads joined in and Scotty organised a seven-a-side match on a field fifty yards by twenty-five, with four-gallon drums marking the boundaries and goal posts. Big Karl was the only other bloke who had played the game, so Scotty appointed him opposing captain and explained the strange rules to the rugby players. 'No tackling and don't handle the ball.'

It didn't sound like a lot of fun. 'A game for pansies,' Dinny declared. 'Count me out!'

Karl was goalie. Despite a couple of courageous diving saves into the gibber stones and the power of the German's boot, Scotty's speed and deceitful feet soon had his team three up. I was challenging for the ball, racing alongside Scotty, when I was caught in a shoulder and neck hold and slammed headfirst into the gibbers. Stunned, I sat up to see Scotty jerk his T-shirt over his head. Livid with rage, he snarled in his Scotts burr, 'You elbowed me, you bastard. You bloody fouled.'

Rising, I tried to shape up, but staggered sideways on a twisted ankle. Fists clenched, Scotty was coming in fast for the kill when Karl hit him like an express train. Grabbing his man in a bear hug he hurtled ten steps and smashed him against the wall of the boss's hut. Again the big wool roller seized the Scotsman like a rag doll and threw him into the wall, before heaving him high above his head and hurling him to the ground. Thunderstruck, the onlookers stood rooted by the sudden raw violence of the action and Karl's awesome strength. They heard him bellow, 'What you do to my mate, huh? Why you hurt my mate, you bastard?' Half-conscious, Scotty sat up, but made no reply. Karl took a few deep breaths, cooling his rage.

The big German was as surprised as everyone else by his explosive reaction. Turning to the silent watchers, he spread his hands and explained, 'He was a bad boy, a naughty boy.' It sounded so ludicrous even Sour Jack, watching from his hut verandah, spared a grin as he turned away. Dinny said to Karl, 'Crikey, mate! Remind me not to be a naughty boy, won't yer, before yer do yer lolly.'

The soccer game was over. Men began to move again, and Tall Snowy and Dinny trotted over to check Scotty for life. Refusing an arm, he rose shakily and retreated to his room. Karl supported me to the hut verandah; somebody fetched the first-aid kit from the shearing shed, and Tall Snowy expertly splinted my ankle. Strapped each day like a race horse, I toiled painfully through the duration of the shed while Karl proved to be a fair-dinkum mate by carrying the lion's share of work for half the pay. The weak ankle was to trouble me for years – always reminding me of the foul-tempered Scotsman.

The rules were lights out and radios off by ten o'clock, but most of the rooms were dark and silent by eight o'clock, or nine o'clock at the latest, as the work-weary team slipped into slumber. The exception was a room shared by Bluey and Tall Snowy, where a game of poker thrived till eleven o'clock some nights. Despite being reprimanded for disturbing the sleep of the occupants of the neighbouring rooms, the games continued. Young Bluey was too weary to play poker; instead he grumbled and growled in the smoky room as he tried to sleep. 'Say your prayers and cuddle your teddy, sonny' and 'Bysy-wysy, little man', was all the sympathy he got. Next door, through the thin dividing wall, Karl and I would hear Dinny growl when he lost, 'Shut up and hide yer head under the blankets, Bluey. Yer putting the mox on me; buggering my concentration!' We got little sleep until the game finished.

Bluey had an idea, and a quiet word with Karl and me. Next day, instead of knocking off early to be first to the showers, we waited for the rousie to finish his chores. We three conspirators then headed half a mile along a dry gully in search of wild pigs.

We knew that each evening a large spotted sow led some twenty grunters of various colours and sizes from their cover in a lignum patch. Pig-chatting in grunts and squeals about the coming feast, they trotted along a well-worn path which led to where the yardman tipped the scraps from the kitchen bin, which was known as a tumbling-tommy.

We hunters had no pig-dog, I was ankle strapped and limpy, and my companions knew sweet Fanny Adams about catching wild pigs. But we were fit, as keen as butcher's blades and primed for vengeance on the poker players.

Pigs have an ordinary sense of smell and vision, but they hear keenly, so Bluey and I crouched behind mimosa bushes a few yards on the down-wind side of the track, with Karl about fifteen yards forward of us. As soon as the pigs had jogged merrily past Bluey and me, Karl jumped into the boss sow's path, threw his arms up and shouted, 'Achtung, you bastards!' Ambushed, the battalion of grunters halted and wheeled in one move. With a chorus of startled squeals, they split in all directions – but by then Bluey was among them. Sprinting and diving he took a young boar in a flying tackle, and they rolled in a cacophonic bundle of bad language and squeals and dust and gibbers until I grabbed a hind leg and hitched three legs together with a leather thong. Karl pulled a flour sack over the head, and the captive instantly quit snapping and squealing, allowing Karl to heft the 120-pound porker onto his shoulder and march for the woolshed.

'Talk about a Clive-bloody-Churchill tackle, Bluey! You're a little ripper!' I exclaimed, referring to the legendary New South Wales rugby league fullback and Kangaroos captain. Still chuckling, we locked the boar in the bottom box of the wool press. Reluctantly, Bluey went to bed in his room beside the poker players after tea to allay suspicion. Karl carried the captive boar to the rear of the huts at nine o'clock, and I swiftly loosed the legs and raised the swing-out window so that Karl could jerk the flour sack free and hurl the porker through the open window.

'Three ladies and I'll look –' said Dinny, announcing three queens and covering a bet as the flying pig hit the middle of the table. Simultaneously, the window slammed shut, the carbide light went out, and a hullabaloo of squeals and grunts and yells and foul language exploded as brave men heroically scrambled for a safe spot on the beds while the panicked porker charged and crashed about in the dark.

A wild yell came with the classic Aussie query: 'What the f—?'

'Get a torch!' and 'Strike a bloody light!' and 'Where's the effing door?' echoed in the pitch black, smoky cell.

Choking back laughter, we pranksters skeltered. We had joined the sleepy 'What the bloody hell's going on?' mob gathering on the verandah, when Darkie flung the door open from outside. Seeing light, the pig and five men rushed for safety, bowling him over. The porker vanished into the night.

Over the following days, the 'flying pig' incident raised plenty of laughs. Fingers were pointed, and on Saturday morning a mock trial was held on the verandah. Tall Snowy felt qualified to sit on the case – word was he'd had spells in Boggo Road Gaol for breaking and entering, so he knew about the legal system. He had twinkling blue eyes and was as friendly and innocent looking as the family Labrador.

'I've fronted more beaks than you'd find on a cackle-nut farm,' he declared, donning a stony visage and mop-head wig.

Accusations flew thick and fast, but all suspects had an alibi – albeit some seemed as absurd as flying pigs. David pointed out that in medieval times pigs and dogs and cats had been put on trial, and even hung, drawn and quartered. He might have been only a learner shearer (a small cut above a rousie in the shed hierarchy) but the boys listened respectfully, for his defeat of Snowy had given his team status considerable weight – although what sprinting ability had to do with British legal history no one thought to inquire. Frankie's younger brother, Peter, a fat, laughing learner shearer, was drafted to play the absent pig – accused of breaking and entering, disturbing the peace and flying without a licence.

David, a natural straight-faced comic, insisted on representing the pig as interpreter and advocate. Peter the pig grunted and squealed his plea and David interpreted, while the boys laughed and whistled and chiacked, yelling 'Hang the bastard – he's as guilty as hell' and 'Draw and quarter him, too – whatever that means'.

The judge's claw-hammer gavel rapped, and David said solemnly, 'My client pleads Not Guilty! Respectfully, your Worship, it is a scientific fact that the familia porcine is flightless; and although a common folklore saying is "and pigs might fly" this is merely sarcasm to admonish fools and liars.'

The magistrate added to his authority by taking time, doodling with his pen and blowing smoke rings. At last, heeding calls of 'Get on with it or we'll hang *you*', he called for silence and spoke of the gravity and importance of the case, before delivering the verdict. 'I declare, by the authority vested in me by this kangaroo court, the accused pig is free to go. The caper of the flying pig was an Act of God. The accused lost his way and flew through the window by chance. "Pigs might fly!" people often say, but now you know – for witness has been sworn – that pigs do fly.

'The fact is that a lot of things are right that folks say aren't true. For instance, Old Snowy here has told us that when he was a boy he was known as George Washington because he could not tell a lie. We know him as a trustworthy gentleman who has seen more yowies and bunyips than you can poke a stick at; and he once ran down that mysterious min min light seen occasionally in outback Queensland and put it in a pickle jar. Who are we to doubt his word?

'Darkie says that when he was a rousie here at Malboona in 1938 he shot a giant rock python. There was the skeleton of a rousie inside. The rousie had been lost six months earlier at crutching time, but they knew it was him because in one bony hand he held a tar-pot and a broom in the other. Recently, the stove had a few words to say – so a flying pig should come as no surprise. The case is dismissed.' Flapping his 'wings' and grunting and squealing while urged on by the jury, Peter the pig celebrated his release by flying the length of the hut verandah.

Knobby Clark was the Malboona cook (or babbler). Pretty well all shearers' cooks wore nicknames, whether they tolerated them or not. His was 'The Busted Oven' – respectfully abbreviated to 'B-O' by the young rousies when out of earshot.

Shearers' cooks were outsiders whose work hours excluded them from the general bonding of a shearing team. To satisfy hungry men with three meals and two smokos per day on working days, they had to rise at 4.30am and put breakfast on at 6.45am. They wouldn't finish in the kitchen before 9.30pm. Come Saturdays and Sundays, however, they would enjoy a sleep-in till 5.30am, as breakfast wasn't till 8am. Poor cooks had short careers, while good babblers, no matter how moody, were always sought, despite their idiosyncrasies.

The talking stove incident had come about because Knobby rarely conversed with people, but often had a heart-to-heart yarn with his kitchen furnishings. Pushing in among the diners he would set to work with a dishcloth, while scolding: 'What's the meaning of this, Table? Only this morning, I wiped you as clean as the soul of a newborn babe, and here you go again – covered in crumbs and tea slop. Won't you ever learn?' He would finish off by whacking the tabletop a couple of times with the cloth, and then proceed to admonish a kerosene fridge, which was issuing smoke through the flue because the wick needed trimming. 'What's this bloody smoking caper, Fridge? No smoking in my kitchen, Fridge! I'll trim your wick quick bloody smart!'

Knobby would deliver these reprimands without the shadow of a smile, oblivious of blokes who might be sitting or standing around. They would choke their chuckles and swallow smiles, for this babbler was known to have the verbal bite of a bailed-up bull terrier and they didn't want to upset a good cook.

One morning at 4.30 Knobby clumped into the kitchen and threw open the fire-box on the big range. Stirring the coals vigorously with a poker he grumbled, 'Good morning, Stove! And what have you got to say for yourself? I stoked you at one o'clock and now, you lazy blighter, you're half out. And what are we going to feed these thankless bastards this morning, Stove? Tell me that!'

The stove replied instantly, 'Give 'em the same shit you usually do, Knobby. And if you jab me with that poker again I'll stuff it right up your ring.' Knobby stumbled backwards. He wheeled about, bolted breakneck through the door and ran twenty yards into the frosty night before it struck him that stoves can't talk. Meanwhile, Dinny slipped from behind the stove and rejoined Bluey and Tall Snowy, who were in on the joke.

Before the breakfast bell rang the story had spread through the camp. There were a couple of sly grins and choked chuckles over breakfast, but nobody made any smart-alec remarks: the odds were that the Busted Oven would blow his stack and pull out if some drongo made a gig of him in public. However, it was later often related that after being chastised by the big Malboona stove, Knobby rarely addressed a fridge or table, or even reprimanded a messy sauce bottle – and he never spoke to a stove at all.

The caper was still drawing cackles of laughter as the team relaxed at smoko that morning. Darkie puckered a devil's grin as he suspiciously sipped his tea. Gazing at the joking culprits, still smugly basking in attention, he drawled casually, 'Old Knobby's no one's clown, yer know. 'Course, I wouldn't give any babbler a bad name by saying he might o' pissed in the tea urn to get square – but all the same, this brew is bloody bitter. Maybe it's only Epsom salts. I happen to know old Knobby always carries a couple o' packets o' salts for special occasions – like if someone don't pay proper respect. If yer get the runs yer'll know it's Epsom salts, otherwise draw yer own conclusions.' He threw another dessert spoon of sugar into his pannikin, and advised, 'If yer double up on the sugar yer'll find it kills the bitter taste.'

A unique drama added to Snowy Hales' legend on the shearing board one morning. Fourteen shearers were bent over, toiling their way through the flock of big, rugged merino rams. The two pressers were filling in time playing chess while waiting for wool, when chaos, punctuated by the waning thumps of the diesel motor, erupted on the board and summoned their curiosity. Along the line shearers shoved half-shorn rams back into the catching pens and joined the gathering workers mesmerised by the scene unfolding on Snowy's number-one stand.

Like many overseers, Sack 'em Jack Butler was a lonely man. Unlike his namesake, the original Sack 'em Richie Jack, he was devoid of self-deprecating humour (or humour of any kind) and made no attempt to form an acquaintance with any of the team. If laughter and camaraderie had ever graced his character they had long been banished by his sour attitude and the slings and arrows of outrageous fortune. David referred to him as 'Cassius of the lean and hungry look'; and if the nom de plume of Sack 'em wasn't enough to condemn him in the court of the working class, the rumour that he had been overseer on the notorious Aboriginal concentration camp of Palm Island put him beyond redemption. But whatever the truth behind his reputation, his courage couldn't be doubted as he stood chest to chest with a shirtless, sweating, bag-booted

Snowy, swapping angry, embittered words before the goggle-eyed team.

Snowy had tangled with an angry ram – one of those rare beasts which kick, fight and bite – in a death but no surrender struggle from go to whoa. A man of iron self-discipline, who never struck a sheep in anger or frustration and rarely used profanity, the gun shearer had spontaneously done his block and with one incredible heave hefted 200 pounds of half-shorn ram arm's length above his head and suspended the monster upside down along the spinning shaft of the running gear.

While the onlookers were astonished and amused, the boss-of-the-board didn't see the funny side. Sour Jack rushed up in a portly trot, shouting, 'You're sacked, sacked, sacked, you bloody madman!'

Eyeballing the overseer from six inches, the shearer fired back, 'And you're a cranky old fool. Run a shearing team? You couldn't run guts for a slow butcher. I'm not sacked, I quit! You silly old bugger!'

'Get that ram down! Get him down, Hales. Get him down at once!' Sack 'em squawked repeatedly. But Snowy never acknowledged him. Turning his back he stuffed his combs and cutters into his tool bag and left the shed.

Snowy had heaved the large ram aloft and lodged him in three seconds flat, but it took a couple of station hands aided by young rousies fifteen minutes (and two forty-four-gallon drums and a plank to stand on) to extricate and lower the ram, unhurt.

Snowy had been a popular, larger-than-life figure and the team's spirit was depleted by his departure. While the shearing continued without a hitch through to cut-out,

Snowy's mighty lift of the big ram and his dramatic exit entered shearing folklore; and for years afterwards blokes who hadn't been within cooee of Malboona shed claimed to have witnessed Snowy's unique feat.

4

MAD JACK AND THE BIRD OF PASSAGE

Following the Malboona cut-out the team was split. I pressed in four- and six-standers until Provo Bob presented an opportunity to test my ability pressing for eight shearers at Oxton Downs, near Julia Creek.

Our cook, who turned out to be a champion, went by the name of Jack. He wore a cherubic smile or satanic smirk as he chose, and he was as volatile as a carbide light – likely to break into wild laughter and a crazy polka while his chops sizzled, or joke of 'ladies of the night' he had 'managed' and the capers of crooked race-horse trainers he had touted for. The boys dubbed him 'Mad Jack'.

The shed flowed peacefully until the third week. Carrying the empty tea urn and singing cheerfully, I strode into the mess for dinner. As I entered through the gauze door the babbler snapped, 'Can't you bring the f—in' tea-billy earlier, Presser? How am I expected to make the tea on time when you're f—in' bludging on the job.'

The outburst shocked me. I explained vehemently that I was paid to bale wool, not to trot the tea urn to and from the shed, but it was the wrong answer. As I sat down to tuck into a hearty feed of roast mutton and veggies, Mad Jack crept up behind me, let go a blood-curdling scream and drove a vegetable chopper into the pine table by my plate. Growling like an enraged bear, Mad Jack jerked the chopper free as I dived under the table and galloped on all fours through the gauze door.

'The cook's gone bloody troppo. He's got a knife and he'll use it!' I warned the approaching shearers and rousies. Mad Jack confirmed my assessment by yelling murderous threats and performing a whooping Indian war dance in front of the kitchen door.

The rep, Allan Hundy, was a game ex-digger who still suffered from World War II wounds. He advanced cautiously with the overseer, Bob Kelly, while Mad Jack carved the air into sections. Bob was one of nature's gentlemen. Trying to pacify him, he suggested Jack drop the knife and talk things over as friends. In reply, Jack advised, 'Youse can all go and get well f—ed with the rough end of a pineapple. And get another cook, too. I won't cook for you crawling mongrels.'

We were glad to hear it.

It was against shed protocol to dob anyone into the police for anything short of murder, so a taxi was phoned and the overseer handed Mad Jack his cheque from the safety of arm's length. Jack's mood changed amazingly once he collared his cheque and boarded the taxi. Waving through the window, he shouted a laughing, 'Hooray, suckers!' as the taxi sped away. Realising that we'd been conned and that the babbler hadn't suddenly suffered a mental breakdown, the team sent him on his way with obscene salutes and various vulgar opinions of his character. Evidence was soon found that the babbler had quaffed Bob's bottle of sherry and two bottles of lemon essence – enough to send anyone mad.

A few years later Jack surfaced around Longreach. Spotting him, I proffered a grin and a hand. Jack denied we had ever met, but he shook hands with a wink and said, 'The name's Joe. Have I got a twin brother?' He informed the team, 'I've been a commission agent for ladies and bookies of good reputation, and have advised punters and would-be smarties at race tracks for a few years.'

The word went around that 'Joe' was living with a notorious murderess who had served a life sentence for conspiring to kill her husband. Naturally, the relationship was a point of gossip about the town and the butt of a few jokes. Joe's red-headed paramour worked as a barmaid and at first she drew more curious customers than she frightened off, but the regulars quickly made acquaintance, appreciated her pleasant efficiency and agreed she had paid her price and should be given a fair go.

Joe and I worked a few ten-standers together. Cooking for twenty-odd men was a big job, but Joe cruised through, sharp of tongue but in good humour. Once in a while he'd whisper with a satanical grin, 'I left my tablets in town, Presser, an' I feel another turn coming on. Anyway, my friend, these zombies need a charge to wake 'em up. Look out when old Joe goes off!'

'That so, Joe?' I'd say and would join the babbler's conspiratorial chuckle. But at the mess table I always ensured I had a view of Joe in front of me. And whether the babbler was really on drugs for a mental disorder I never discovered.

After Oxton Downs cut-out Provo Bob employed me as a piece-picker and reserve wool roller at Carandotta for sixteen shearers, so that a back-up presser would be on hand if one of the two pressers broke down.

'Where's Carandotta?' I asked Darkie.

'Two miles west of sunset, son,' Darkie replied as soberly as a doctor delivering a death sentence.

'Are you going?' I asked hopefully.

'No bloody fear! I was there twice in 1938 – the first time and the last time. In fact, I was the only survivor. The wild blacks speared seven blokes for chasin' their ladies, and the babbler – "Poisoner" Johnson – did for the rest with hoop snake poison. Believe me, son. I was there!'

The boss-of-the-board, Charley Thompson, and Tom Murphy, his brother-in-law, a hardy presser who still sported his World War II flying officer moustache, drove the 600 miles from their home town, Charters Towers. Travelling with them were the babbler, the Busted Oven and the classer, Arthur Cox, known behind his back by respectful rouseabouts as Old Dry Balls.

With the bulk of the team I boarded the *Inlander*, the new air-conditioned diesel train running from Townsville to Mount Isa. Provo Bob had laid down the law that no grog was to go to Carandotta. It was going to be a long time between drinks, so some of the boys got stuck into the booze. The roisterers rambled the length of the train, stirring the girls in the service car and drinking the cold room dry before they were halfway to Cloncurry. There, we threw their swags and ports on to a semi-trailer and climbed aboard to rough-ride the tray for 170 rugged miles to Carandotta.

The six weeks at the big shed on the drab plain dragged. At weekends we lightened our boredom with fishing trips to the Georgina River, cricket on the claypan, and the usual card games, boxing, betting on the races, draughts, chess and two-up. One Sunday morning as I stood looking at the remains of a burnt tank stand, a bow-legged, leathery station hand interrupted my pondering. 'I was here when she burnt, mate.'

I reckoned the ringer looked old and brown enough to be an original de Satge offspring and asked what happened.

'It was the babblin' brook done it. A cove they called The Bird of Passage. He made the biggest billy of tea ever brewed.'

'Yeah, I've heard of him,' I said.

'A top babbler, an' sure to fly if anyone was silly enough to hurt his feelings — which was easier than catchin' cold. Anyway, him and the slushy have been cookin' for sixty men for four weeks and no trouble, when the babbler gets a complaint that the tea is too weak. The Bird of Passage says, "I've been makin' the same brew for the crying bastards every day. Strong tea they want! I'll give 'em strong tea."

'The boys haven't seen a pub or a woman for over a month, and they're gettin' stale and cranky. A couple of days later the rep and committee man have to walk from the shed to the kitchen to chip The Bird again. They're not happy, but the team has made the bullet and the rep has to fire the gun. They tell The Bird that there wasn't enough tea to go around at smoko. "Is that so?" says The Bird.

'Just before the dinner bell rings at the shed, a rousie looks out the window and yells, "Fire!"

'They take off like a Melbourne Cup field to save their gear – but it's not the huts burning, it's the tank stand. She's ablaze, and standing beside her is an empty tea chest. The Bird and his offsider have thrown sixty pounds o' billy tea into the tank, shoved half the woodheap underneath and lit her up! They're disappearing in The Bird's T-Model Ford, and on the kitchen table is a note reading, "Send my cheque to Charleville. PS: I hope this is enough tea and it's strong enough, you bastards!"'

'That's a bloody good yarn, old mate,' I enthused.

The ringer placed a work-worn hand over his heart and declared, 'Too bloody right! I was there. Fair dinkum – saw it with me own two eyes!'

Carandotta cut-out at long last and we eagerly climbed on the semi-trailer to return to Cloncurry, where half of the boys tumbled off before she stopped rolling and rushed into the nearest pub. They got half-charged in a hurry, and a few simmering tensions exploded into brief punch-ups, before they boarded the *Inlander*. The locals dropped off at Hughenden, but a few cashed-up shearers rode impatiently to Townsville, where they caught a plane south. I boarded the famous 'Sunlander' train to Brisbane – and a family Christmas.

The family had moved to West End – a spacious Queenslander overlooking Davies Park and the Brisbane

River. There were hugs with Mum and a handshake with Dad, who couldn't disguise his pride that his weedy shiralee had grown into a husky wool presser. Mum was relieved her prayers had delivered her wandering boy home safely. On Christmas Eve we gathered around Dad, who sat in our large lounge room beside a bushy Christmas tree, beneath which a pile of wrapped gifts had been added to for days. He jovially read the labels and handed out the presents to claps and joyous clamour; then it was midnight mass. Come Christmas morning we boys hiked to Davies Park to join a game of cricket while Mum and the girls sweated in the kitchen.

As ever, it was a cracker Christmas feast: Mum said grace, and then came love and laughter and jokes over ham, roast chooks and salad, followed by Mum's scrumptious cold Christmas duff with whipped cream and fruit salad. Then it was washing-up time: the boys' turn in the kitchen.

In the New Year I travelled to Longreach with my brother Barry, and Nellie, a kelpie/collie crossbred sheep dog. Nellie was brick red with alert ears and a white tip to her tail. On the road droving she could be relied on to silently steady the leading sheep, work on the wing or hurry the tail along. Penning up in the shed she would run over the backs of woollies barking her commands. As a pup she had come to me five years earlier when I was a station hand on Yarrawonga station, St George, and I had learnt to love her as only a lonely boy can love a dog.

I had a new green Volkswagen, and Barry, who believed he was world champion race car driver Jack Brabham, took the wheel over most of the journey. He was bound for Jundah to take up a job with the post office and to board with the sergeant of police Jack Williams and his wife, Betty – long-time family friends. Barry had arranged to catch the Jundah mail at the Longreach post office about sundown, so we left Brisbane at 4am to drive 700 miles. Beyond Roma the so-called highway was mostly dirt road, but we arrived in time to have a cafe meal and find Max Cramb loading his three-ton truck for the 140-mile overnight run to Jundah. Barry got into the truck and gave me a wave, before heading off with Max.

I pointed the VW towards Hughenden, 210 miles north, and camped beside the road.

Early 1959 found me pressing and penning-up for six crutchers who were shearing the wool around the sheep's tails to inhibit blowfly strike, at Aberfoyle station. There Nellie whelped nine pups. I gave eight away after selecting a stumpy-tailed blue and black pup I named Zulu. He was the image of his cattle-dog sire, a Smithfield/blue heeler cross.

After Aberfoyle, sixty miles south of Torrens Creek, I drove 400 miles to Lucknow, sixty miles from Boulia, and then 300 miles to a run of sheds in the Longreach district. No wonder shearers said, 'Work with UNGRA to see the world.'

After returning to my family in Brisbane for the mid-year break at West End, Dad and I drove back to Hughenden for the second half of the year. On the way we called at Boomaroo station, owned by Frank Morrison, an old mate of Dad's. Frank saw Nellie at work in the sheep yards and offered twenty-five quid for her, a good home and the work she loved. I was sorry to part with her, but I knew that two dogs would be difficult for me to handle and reluctantly let her go.

Dad and I worked in separate runs until November, when we joined up at Tarbrax station, south of Gilliat, before completing the run at Oxton Downs. This time the babbler was sane – as shearers' cooks go. But Fred wasn't in Mad Jack's class as a cook or entertainer.

A big jovial bloke of middle years, he claimed to have been a Brisbane boxing trainer, and took me in hand for a six-round main event in Julia Creek. I had always boxed off the back foot, but my new trainer insisted, 'A bloke who hits as hard as you do, son, should tear in from the first bell and get it over with.' My opponent, Mike, had the same plan, so the crowd of some 300 got a humdinger of a brawl. Mike had a couple of inches and a stone weight advantage; he was wavy-haired and had a face more akin to a Hollywood hero than a tough bush worker and scrapper.

Tearing in, I nearly did get it over in the first round. I copped some wicked punches early and had to duck and dodge to stay upright. But I got my man's measure in a slugging third round and carried the fourth, staggering him seconds from the bell. The promoter-cum-referee then stopped the fight – two rounds short of the scheduled distance of six rounds – and declared Mike the winner on points. The shearers had backed me with good money. Reckoning their man was finishing the stronger, they raised a ruckus, but the referee held fast, claiming an error had advertised the fight as six rounds instead of four. 'That's bullshit,' the cook bellowed, and issued a challenge for a return bout. Bruised and bloody, Mike wasn't interested; and nursing broken ribs and a battered nose I wasn't sorry, despite my attempt at false bravado: 'Any time, any place.'

5

THE GIRL AND THE HONEYEATER

In the New Year I did a run with Alex Meekin, the legendary Charleville shearing contractor. A team of thirty – including fourteen shearers – signed on at Milo, a historic station near Adavale. Johnny Flett from Newcastle was a good mate on the Ferrier press; Snowy Hales was the gun shearer, followed by Les Goodman, a Charleville champion. As was his wont, Snowy organised weekend sports. I took my gloves to the Adavale pub, where Darcy Delaney and I put on exhibition spars. Luckily for me Darcy didn't possess the skill or power of his older brother, Kevin, who had won the Australian welterweight championship from the great Queensland fighter Tommy Burns.

The first day there I asked the babbler, Stan – known as The Honeyeater – 'Can I swing a tea towel for yer? I've got form! Mum trained us kids young.' Cooking for thirty men with no offsider and close to pension age, Stan was glad of a hand and happily accepted, and we soon became yarning and fishing mates. Bill, a wool roller, was another of Stan's fishing mates and also a willing helper with kitchen chores.

Quiet and inoffensive by nature, Bill had been a quality shearer for thirty years until his failing back finally made him throw in the towel. With a home and a wife and family in Toowoomba, he could have secured a steady job with the famous Southern Cross Foundry or the Toowoomba Council, as many ageing shearers did, but like a lot of bush people he found the noisy constancy of city life depressing. Missing the camaraderie of old mates, and the natural music of the grey-green bush, he had enlisted in Alex Meekin's team as a shedhand.

Busy as he was, Stan found time to read and chat. A precise, religious man, he liked to season a wide general knowledge with a keen interest in sport and politics and a pithy wit. Formerly a good shearer, he was now rated a crackerjack babbler who took pride in the produce and cleanliness of his kitchen and strictly policed his rules. 'Respectable dress in my kitchen' meant shoes and shirts to be worn for all meals; and carnal bad language and any jokes or talk disrespectful to women would see The Honeyeater fluff his feathers and drop on the perpetrator with the fierce righteousness of a wedge-tailed eagle on a careless bunny.

After Milo, eight shearers went to Ray station for eight days' work. Snowy Hales pulled out at cut-out to go opal gouging and a team of six went to a couple more sheds. The team worked smoothly, but Stan was becoming a bit touchy, as babblers often do after a few months in the bush working alone seven days a week from four in the morning till nine at night. One day Bert, a jovial part-Aboriginal shearer, bowled in the door for dinner, calling, 'Hi-ho, Cookie! What's on the menu today, you greasy old bugger?'

It was the kind of boisterous entry he'd often made before, especially on a Friday, when his mood was inflated by the promise of a weekend's rest from grinding sweaty toil, the welcoming arms of his wife, and a few bets and beers on Saturday arvo – followed by Sunday with the family at the footy barracking for the Charleville team.

The babbler usually responded in good part, but now he exploded, 'Don't call me greasy, you clown. A man's job depends on his reputation. I don't go around calling you a snagger.'

The shearer took a pull. 'Sorry, Cookie, just joking . . . I never guessed you were so touchy.'

Flourishing a large vegetable knife like a blood-thirsty Excalibur, Stan shouted, 'Why wouldn't I be touchy cooking for an idiot? And don't call me Cookie! My name's Stan . . . Got it? Stan!'

With his plate loaded, Bert joked as he scuttled to the far side of the table, 'Make room for me, boys, so I can sit down before the babbler cuts me down.'

The jest drew laughter, but the old hands knew better. Back on the board Les Goodman opined, 'It won't be long now before Stanley will be taking a month or so off at his fishing shack on the beach. Great place, Wynnum! Me and the wife have visited The Honeyeater a couple of times. That's the life! Clubs, pubs, no hot weather . . . No bloody sheep! I dunno why he comes back at all. He owns the shack, an' he's got a quid. But he'll be back come July, an' he'll say, "I've got to help old Alex out" – that's always his excuse.'

The shed cut-out on a Wednesday afternoon, and the team was to sign at a neighbouring property on the following Monday. The Charleville men headed for home and Stan went with them, intending to stay with friends. I had phoned ahead and received permission to camp in the neighbour's shearing quarters, where I could enjoy some shooting and fishing while waiting for sign-on.

After pressing-up the last of the clip I drove to Quilpie. Zulu was ensconced in the passenger's seat. Now a year old, he had grown into a playful, humorous mate, a watch dog, an effective penning-up dog and a fierce hunter of roos and wild pigs. He was also a good team player and enjoyed fielding at cricket, even if he often took off with the ball. Over time he had become expert at a man's job which nobody wanted on frosty mornings: swimming across a waterhole dragging a cord attached to a fish net.

I ate a cafe meal of steak, eggs and chips before booking a room at a pub and going to the bar. With a beer in hand I moved to a corner where I could chat up the barmaid, if she was in the mood.

Alice was a tall, slender, sandy-haired girl of nineteen, who had arrived from Brisbane only a week earlier. She was working the 2pm till 10pm shift and was too tired to meet up afterwards, but agreed to take a drive to see the shearing shed on Thursday morning, as long as she returned in time for work.

It was a thirty-mile drive to the property. Zulu was banished to the back seat, and Alice lit up a cigarette. She told me that she was an office worker and she'd left Brisbane because of a broken romance. Seeing the Quilpie job advertised she had decided on the spot that a change of scenery and a new job would be good for her. 'So here I am!'

I pulled in at the homestead – as required by protocol – and introduced myself. As we drove on a mile to the shearers' quarters I was confident I had good reason to feel as happy as the proverbial Larry: an attractive girl sat by my side, and I felt pleasantly optimistic that given time she might be interested in activities more amorous than wandering through a shearing shed.

I was surprised to see Stan coming out of the kitchen to meet me. 'G'day,' he hailed, but when he saw the girl his welcoming smile disappeared. 'What's she doing here?' he asked, ignoring the girl's feelings.

'Alice is with me,' I said shortly, and took her hand. I escorted Alice to the shed without mentioning Stan, and wandered through the sheep yards and the shed, explaining how it all worked.

It was a shearer's tradition to always invite a traveller to share a meal, and though hostile towards the girl Stan still served a lunch of cold meat and salad, maintaining his silence while she and I conversed in a tense atmosphere.

In an effort to defrost the cook's mood, I exclaimed enthusiastically, 'Bewdy, Stan! Fresh lettuce and tomatoes. You're a wizard!' To Alice I said, 'Stan's got a few old girlfriends around the station cooks. They all greet him with a happy smile and fresh veggies.'

Stan snapped, 'They're not girlfriends. And they don't employ a station cook here. The missus gives me veggies every year – from her garden. I've been coming here for four years. I'm respected. We called in here to drop the kitchen gear and some stores off on the way to town. I knew you'd be here, Alan, so I reckoned I'd stay and we could throw a line in and catch a few fish.'

Stan's attitude remained churlish while we ate, and when Alice offered to wash up he barked, 'No need to get your hands dirty, girlie. I'll do it!' Rebuffed, she wandered off along the verandah. I began to follow, intending to apologise for my mate's unexpected behaviour, but Stan rose and grabbed my arm. 'You had no business bringing her out here; she's nothing but a slut. She's been kicked out of a brothel in Brisbane. Jimmy the publican has brought her out. He does that – brings out sluts!'

His attack was so vehement and certain, that I surmised they must have had previous acquaintance. 'Do you know her?' I questioned angrily.

'Never met her. But I can pick 'em. She's diseased – look at the pimples on her face. She's one of Jimmy the publican's sluts. He brings 'em out all the time. They get on their backs and he makes a quid on the side.'

'You dirty old bastard,' I snarled into Stan's face. 'You don't even know her. She's a respectable girl. If anyone else spoke about a woman like that you'd chuck them out of your kitchen. You bloody old hypocrite.'

The clatter of the girl's shoes on the hardwood verandah and the swing of the gauze door truncated the row. 'I want to go back to town. Please take me back to town,' Alice pleaded between sobs, holding a handkerchief to her face. She halted a few feet away. As I looked at the distressed girl in her off-white cotton dress decorated with yellow daisies my anger evaporated, to be replaced by shame for my friend's outburst and empathy with the girl. She was my responsibility.

Alice ran towards the Volkswagen, while Zulu sprinted ahead and leapt through the window. Burrowing her head into a cushion against the door, Alice sobbed, impregnable to my attempts to apologise and explain. 'I'm sorry,' I said. 'I just dunno what came over him. He's a mate of mine – usually a champion bloke.'

I drove off and twenty minutes passed in miserable silence before Alice dried her eyes and sat up. Reaching over the back, she patted the dog, then lit a cigarette and said bitterly, 'Have you got any other friends like that *nice* old man?' I was relieved she was talking again, but the sharp edge of her scalpel found me struggling to find an answer.

'He's jealous, you know,' she said. 'That's what's wrong with the old man.'

'Jealous? Jealous of you and me?' Cocooned by youth's inexperience, I believed that jealousy was sparked only by conflicting desires for men and women of the opposite sex. She gave me a knowing smile in reply.

I said, 'Stan's far too old for you! Two or three times your age! That doesn't make sense.' Twenty years passed before I finally understood. Twenty years later, when life had impaled me on the spiteful spears of jealousy and revenge, the scene revisited me and I realised I had been as thick as a brick: Stan had resented her invasion of his man's world. He was jealous of Alice's claim on the attention of the young man he had taken under his wing – his fishing mate.

We pulled up behind the pub near the staff quarters and I put an arm around her. 'When can I see you again?' I asked quietly.

'I'm working till ten o'clock, but it's not much use. I've decided to go back to Brisbane.'

'Is that about today? About Stan?'

'No. The old man had nothing to do with it.'

It was my turn to be wise. I grinned and exclaimed as brightly as I could, 'You're going back to a boyfriend! I'll betcha guineas to gooseberries.'

She looked at her toes and scuffed her feet before frankly meeting my gaze. 'Maybe, but not yet. I don't like it here. I don't like bar work; people are all over me. I told Jimmy the boss. I gave notice! He's a friend of Dad's and he's got me a lift on a wool-truck – all the way to Brisbane.'

We kissed fondly and, still chafing from the hurt of my row with Stan, I decided to drive on to Charleville rather than spend a tense long weekend in the bush. Returning in time for tea on Sunday, Stan greeted me abrasively. 'If you'd had the decency to tell me you were going to Charleville, Alan, I would have gone with you. You left me stuck out here in the bush like a shag on a rock.'

'Is that so?' I replied coolly.

The exchange was our only conversation for several days apart from chilly courtesies. Yet, we harboured an innate liking for each other. On the Thursday night Stan asked, 'What are your plans for the weekend, Alan? The boys are going to town, and Bill and I are going to throw a line.'

I accepted. It was an offer to let bygones be bygones; an opportunity to fill in a quiet weekend in the bush restoring friendship and energy. I would stoke up the copper early and get the washing done before taking my rifle and Zulu for a pig chase before lunch; and after idling the arvo away there would be a quiet night's fishing with Bill and Stan.

6

THE SHEARING CONTRACTOR

Bill and I went frog catching for bait on Friday night. Bill carried a sturdy green-stick waddy in one hand. He announced, 'I'll go ahead with the torch and watch out for snakes.'

'Keep a good watch, mate,' I advised, grinning in the darkness. 'If a man gets bit by a big mulga snake he's a goner.' I tied up a disappointed Zulu, for snakebite is usually fatal to dogs, and caught up with Bill.

As we were picking our way through lignum bushes bordering a creek I threw a stick between Bill's legs and yelled, 'Look out, Bill! Snake!'

The stick snagged in Bill's trouser cuff. Convinced he'd been fatally fanged and had only minutes to live, Bill swung his waddy and danced the quickstep while the torch beam waltzed wildly through overhead coolibah branches. Nearly speechless with laughter I managed to shout, 'It's only a stick, Bill. You're *alright*! Take it easy, mate!'

Bill tumbled six feet into the dry creek bed, and his yells of terror turned to gasps of pain and a shout of anger: 'You *stupid* bastard, Presser!'

With his arm around my shoulder Bill struggled from the creek and dragged a leg across half a mile of rough ground to the huts. I felt gutted: rough play was fun among fit young blokes, but I'd been careless of Bill's age and infirmity; my thoughtless rowdiness had wrecked a workmate's back. His only complaint, however, was: 'Bloody sciatic nerve is pinched again, Presser. I'll be laid up for two or three weeks.'

Bill related the story of the backfiring joke, and in response Stan dealt me a furious tongue-lashing that I knew I deserved. Solicitously, The Honeyeater made Bill a mug of cocoa, and the stricken wool roller swallowed a couple of painkillers he carried for such emergencies. In the morning he struggled from his stretcher, and I assisted him to a chair in the sunlight. 'Lucky you're a tough old bugger,' I joked half-heartedly.

'Don't feel bad, son. These things happen. But I'll get you to run me to the hospital in Charleville.' He squeezed a grin as he eased into the chair, and said gamely, 'I'm as stiff as a honeymoon prick, Presser. But I'll be right once I warm up.'

Stan was mopping the dining-room floor when I entered nervously, looking for breakfast. 'There's some boiled eggs and bacon,' he briefed me. 'Make yourself some toast and take a feed over to Bill. And then come back and fill the fridges with kerosene. I'm going to town, too.'

We loaded our gear into Bill's FJ Holden sedan. Groaning, he crawled in a rear door and lay down on the seat. I took the wheel and Stan took the window, while Zulu, who waxed fat on his friendship with babblers, sat up grinning between us on the bench seat. He was soon scrambling all over Stan to look out for roos and pigs. Stan took this rough assault on his neat sports coat and trousers surprisingly well. 'Pull over, Alan,' he said. 'If Zulu wants the window seat that much he can have it, and I'll sit in the middle.'

When we regained top gear, Stan observed, 'It's one of life's great mysteries: if you blow in a dog's face he'll panic, but put him in a car and straight away he'll stick his head out the window and grin and bark happiness into a fifty-mile-an-hour gale.'

I began to relax, for The Honeyeater was returning to his habitual informative feisty character. Coming into Quilpie he said, 'Let me off at the post office, Alan. I'll ring Alex and tell him he'll need a shedhand and a cook, while you fill 'er up.'

Quilpie to Charleville was a punishing drive over a rough-as-guts dirt road, and cushioning Bill's back as I negotiated bone-jarring potholes, grids and corrugations made the journey much slower than usual. Bill suffered without complaint, save shouting a couple of times when a spring bottomed, 'You're a worse bloody driver than Alex Meekin, Presser.'

After leaving Bill at the hospital, I dropped Stan at his mate's place and drove to Alex Meekin's home.

*

THE SHEARING CONTRACTOR

It was close to three o'clock when I ran up the worn front stairs of Meekin's old Queenslander. On the verandah I slowed down at the open double door to his office.

Alex Meekin's good name was a byword wherever shearers yarned – from Hamilton in Victoria to Cloncurry in north-west Queensland, from Broken Hill to Goulburn and Longreach.

I had enjoyed the run with the fabled contractor. I'd earned good money and found fair dinkum workmates; and through yarning and listening I had formed the impression that the Old Boss saw himself as a father figure; a rough-around-the-edges mentor of worthwhile young men seeking a future in the uncertainty of the shearing industry.

During the 1956 shearers' strike the contractor had stuck with his old hands and the Union. Chuckling gruffly, he would recall, 'Before the fifty-six strike I was the biggest private shearing contractor in Australia. Now I'm the smallest.' In fact, his teams still shore several hundred thousand sheep per year.

The contractor had just paid off all but two stragglers of one of his shearing teams. I waited by the open doors and heard the young brothers in warm debate with Alex. 'Can't you place us together, Mr Meekin?' George, the younger brother asked firmly.

Alex, who was seventy-two, was sitting down behind a large old desk, occupying the only chair. Although he usually had respect for the men he employed he didn't want them dropping into a chair to debate or argue with him at length. Speaking gruffly in the loud voice of the hard-of-hearing, he insisted, 'As I said, George, I've placed you wool rolling starting Monday, but I haven't got a possie in that team – or any other team – for a learner shearer like your brother for a fortnight. You know it's always chancy until the end of the financial year. From July I can keep both of you going till the end of November.'

'You promised to keep us together and in work,' Frank accused. 'And I'm not a learner! I've had a pen for twelve months – and I shear over a hundred a day.'

'And I've kept you together for over three months, Frank,' the contractor shouted, his short wick flaring at the whippersnapper's implication. 'Don't tell me my word's not good!'

The Old Boss was proud of his reputation for straight talking compared with contractors who stretched the truth to keep men handily on call while work was slack. Such men earned appellatives such as 'Promise me Perc', 'The Liar Bird', 'Bullshit Bill' and the 'Artful Dodger'.

The ferocity of the contractor's bellow knocked the brothers back a step and silenced them, while on the verandah I was awed by the power of his personality. Pushing a thick foolscap ledger and two out-size cheques over the desk, Alex said gruffly, 'Sign this.' He watched them sign and carefully fold and pocket their cheques and statements. Then he said with a hint of conciliation, 'I'll tell you what I can offer. I can make a switch: send you out together rouseabouting for two weeks; then Frank, you'll be back shearing. I've got four teams starting the first week in July and more work than you can poke a stick at.'

The young men glanced at each other, and then Frank said, 'No thanks. While we're up this way we'll have a look at the country up north. We've got to be home in early

THE SHEARING CONTRACTOR

October to do Dad's shearing – he's got twelve hundred ewes and lambs.'

Alex looked up. 'Twelve hundred, huh?! Do you call them by their first names?'

Ignoring the sarcasm, Frank went on firmly, 'There's plenty of work around home then – around Grenfell and Forbes. We'll leave for Hughenden after lunch.'

I anticipated another explosion, but the contractor began sorting papers as he harrumphed. 'Hughenden, huh! See Clever Claude or Bullshit Bill – they'll fix you up.'

The brothers hesitated – they were hard-working, respectful lads, products of a conservative home and community, whose experience in the industry was drawn from working for neighbouring sheep cockies. They turned to each other for support before George said, 'Ah, Mr Meekin, we appreciate what you've done for us. Could you write us out a reference please? Just say we're reliable, good workers – and sober. It might help us get a pen.'

I could have sworn I saw the flicker of a predatory smile cross the contractor's craggy visage – like the smirk of a resting cat spying a fat unwary quail.

'A reference!' he exclaimed, feigning astonishment. 'You've already got one. Look in your pocket!' The shearer drew the stapled cheque and statement from his pocket while Alex watched him keenly. Fascinated, I followed the proceedings – two or three times I'd heard the tale of Alex and *the reference* related over a campfire. I'd assumed it was a classic bit of outback folklore inspired by a campfire yarn-spinner. Now I saw it right in front of me, as George, in slow motion it seemed, unfolded his cheque and statement. He examined it and said, 'I don't see any reference, Mr Meekin.'

'Take a closer look, boy,' Alex said gruffly. 'You see the camel train in the corner? Slow and rough – like your shearing. You've got your reference, lad. It's in your hand.'

Any argument left in the lads evaporated when Alex Meekin rose to his full height. At six feet two inches and thirteen stone of fatless bone and sinew, he had features as rugged and weathered as a Mount Rushmore president's stony face. The chastened brothers brushed past me and retreated down the stairs in silence. Reaching their Holden FJ ute, they cast a resentful glance at Alex who had joined me on the verandah. The contractor acknowledged it with a casual salute and a self-satisfied chuckle.

I followed Alex into his office. He resumed his seat of authority and declared, 'You're Jack Blunt's son. Jack shore with me before the war. Good shearer; good man. Where is he now?'

'Dad's camp cook for a Regional Electricity Board gang in the south-east. They work close to Brisbane. He's home with Mum and the family on weekends. He reckons this time he's given the bog-eye away for good.'

Alex harrumphed acknowledgement. 'The classer tells me you're doing a good job. I like that – a young fella who's not afraid of hard work. You go to Pinkilla in June. Eight shearers. Can you handle it?'

'Yeah, 'course I can. I pressed for eight at Nerrigundah. On the Koertz,' I added referring to a common make of wool press which most good pressers rated inferior to the Ferrier. 'I pressed thirty-eight bales for two days, and then pressed forty-seven on cut-out day. But I had to work a bit of overtime.'

'There's more wool at Pinkilla. If it's too much I'll give

you an offsider. You don't want to bust yourself. A week's break after Pinkilla, then I can look after you till December. Do you want it?'

Alex patiently jotted figures while I compared my options. Over the previous two years I had established a pressing run with the United Grazier's Shearing Co-op in Hughenden, signing on from Torrens Creek to the Georgina. Bill Blackwood, known around the traps as Bullshit Bill, had taken over as manager from Provo Bob and had guaranteed me 'big mobs of work' from early July till the end of November. Still, I liked the Charleville situation. I'd made a few friends, and a couple of married shearers had taken me home to 'meet the wife and kids and have a feed', divining that I would enjoy the warmth of family. Plus there was fine camping and fishing for Murray cod and yellowbelly along the shady reaches of the Warrego River. The mighty cod weren't found further north in the Flinders River or the Lake Eyre basin catchment; in fact, bountiful fish holes were few and far between across the north-west.

The fishing was a big lure, but in truth it was the much larger lure of attractive, lively young women that helped me reach a decision. 'That sounds good, Alex,' I said. 'Thanks. I'm with you.'

'You won't be sorry, lad. Stick with me. I'll look after you.' He waved a gnarly hand to indicate the Holden I had parked in the street and added, 'That's old Bill's car you've got there. How is old Bill?' Old Bill was about fifteen years Alex's junior, but Alex seemed unaware.

'Pretty crook. His back's buggered. They carried him into the hospital on a stretcher. He said to leave his car here.'

The contractor said, 'You'll need a lift back to the shed. I'll pick you up at one o'clock tomorrow.' It was more of an order than an offer.

'Have we got a cook?' I asked.

Alex fixed me with a look that signalled I had asked a silly question, before answering brusquely, 'Good cook. Bloody good cook! You've heard of him: The Bald Eagle. The Eagle is going to the races today. The classer will take him out in the morning.'

The following morning I was eating breakfast at the Charleville Hotel when a skinny blond bloke of about twenty-five breezed in, grinning. His neatly styled clothes matched his confidence: desert boots and long yellow socks, striped casual shorts and a collared T-shirt. 'Mind if I join you?' he opened, removing his panama hat. 'I'm Jimmy G – from Sydney.'

'They call me Presser, or AJ, or Alan – take yer pick,' I said. Seeing a chance I couldn't let slip, I bunged on a Dad and Dave drawl to upstage Jimmy's clipped Sydney accent. 'Fair dinkum! Sydney, hey! The old steak and kidney. Been there a few times meself, yer know. Bloody good town, too! I stayed at the pub. By the way, *who* is running the pub in Sydney now?'

It was a hoary gambit; every bushie had heard it, but Jimmy G fell for it and, determined not to embarrass his new friend, said, 'Well, ah, there's more than one pub in Sydney these days, AJ. It must be quite a while since you were there.'

'Too right, mate,' I drawled. 'Now, let me see . . . I was going to Melbourne to see the Cup, and I dropped into Sydney on the way. But I reckon it was the year that Phar Lap won the Cup, 1930 – or was it Carbine in 1890?'

Realising he'd been hooked, Jimmy laughed wholeheartedly with me.

We yarned and joked over breakfast, and discovered we were going to the same shed, before Jimmy became serious. 'Would you do me a favour, AJ? Could I get my mail addressed to you?'

'Well, sure,' I replied, 'but I just get mine sent care of Alex Meekin, Shearing Contractor. The mail truck delivers it or someone picks it up.'

'No. I mean, if it's put inside a large envelope with your name on it?'

I tried not to show my surprise. 'Yeah, that'll be right – but what's the caper? Did you knock off a bank or get caught with the boss's missus, or something?'

'No, nothing like that,' Jimmy protested. 'I'm not wanted by the cops. There was a big warehouse robbery. I was caretaker, but I had nothing to do with it.'

The industry hid a few blokes on the run, so I was tempted to say *A likely story, old mate!* But I bit my tongue.

Sharp at one o'clock Alex picked Jimmy, Zulu and me up and we loaded the well-type Fargo ute with stores. With Zulu rushing madly around on the covering tarp barking triumphant abuse at the town's stray mongrels, we headed for Quilpie.

Around the shearing sheds I had heard firsthand evidence that Alex was the worst driver in the back country. It was said that he put his truck into second gear and kept it there till journey's end, and he never missed a bump if he could help it. 'The only time I've seen him change gear,' the red-combed 'Rooster' Kelly declared, 'is when he misses a bump and has to back up to have another crack at it. The fact is he's punished so many axle-busters over the years that now when they hear Alex comin' they go bush – but the Old Boss chases them into the mulga and runs them down.'

I had assumed that most of the yarns about Alex's awful driving had sprung from the exaggerations of pull-up-a-stump comedians, but after a few miles of bouncing about the cabin, sitting between Alex and Jimmy, I knew that I would swear the tales were gospel – if I survived to bear witness. 'Fair dinkum, fellas,' I silently rehearsed to keep my mind off the Grim Reaper's proximity, 'old Alex is the *best* driver I've come across. He's the only bloke I know who can steer a truck through a ten-foot grid sideways at forty miles an hour. But my uncle Kev gives the nod to Jack Williams – he's the police sergeant at Jundah now. They cut sleepers for the railway around Blackbutt in the thirties. With the dray loaded with sleepers, Jack would gallop Josie, his Clydesdale mare, down the mountain slope shaving bark off the trees on both sides with the wheel hubs.'

I was jerked back to the here and now by my head banging against the roof, and Jimmy landing on my lap simultaneously. Hunched over the wheel, Alex held forty-five miles an hour in second gear and drove with a grip of steel as the truck slew wildly out of bull-dust holes, through grids, danced across corrugations and charged through dry creek beds where the concrete causeway had long since broken into chunks. Luckily, traffic was rare because vision through the red dust haze was at best 300 yards, and dropped instantly to a frightening zero as we plunged into a pall left by a passing vehicle. Jimmy and I braced ourselves, placing feet on the firewall and palms on the dash. Even so we cannoned into each other, as a cloud of fine dust coated every crevice of our heads and bodies. Even Zulu, the erstwhile King of Utes who was usually so fearless of falling, took refuge deep in the load.

Miraculously, we arrived at our destination right side up. Early the next morning the weather was chilly, but following my regular practice I warmed up with a mile run before heading for the showers. In the washhouse I witnessed another of the legend's idiosyncrasies: with his face lathered with a coarse laundry soap we called 'Kerosene Bouquet', Alex was using the rear side of a tin dish for a mirror while he shaved with a venerable bone-handled cut-throat razor.

'That sounds like a billy-goat pissing on tin, Alex,' Les Goodman said, laughing as he swilled his face with warm water from a copper that had been stoked overnight.

Alex grunted while he scraped. 'The presser's mad,' he muttered as I turned on the tap for a cold shower. 'Who does he think he is? Herb Elliott?'

'Too right he's mad! You're both bloody mad,' Les replied.

7

JIMMY G AND THE BALD EAGLE

As usual it was lights out and radios off at ten o'clock. But after an exhausting day's hard work followed by a longneck or two, many would be asleep by eight o'clock. Those who stayed up relaxed with the girly delights of *Man Magazine* and Carter Brown thrillers, or reading anything from racing form guides to *Australasian Post*, the *Reader's Digest*, and popular novels and comics.

I had scored a room on my own because I read till late, escaping the confines of the shearers' huts to be in the company of AJ Cronin, John Cleary, Steinbeck and Tolstoy. Yet I invited Jimmy to share the room, guessing the young Sydneysider would be good company with his laughing, sociable outlook and experience of city life. He didn't take long to prove himself a trump, telling entertaining stories that didn't reveal much about himself while disclosing a resume of night work as a taxi driver, night-club waiter, barman and night watchman. He dug into my novels, but passed a lot of time lying on his back lost in thought. Waking in the small hours, I often saw the tip of his cigarette glowing; and sometimes saw him put on a coat, pick up a torch and go walking.

Jimmy began his apprenticeship on the board learning to pick up fleeces and throw them on the wool roller's table. He was dubbed Wait-a-While when Les Goodman commanded, 'Wool away! Liven up, lad! Get this bloody fleece out of my way.'

'Wait a while *yourself*, will you! Can't you see I'm busy?' yelled Jimmy, who was struggling to sort out another fleece and didn't know that picker-ups were expected to step lively to a shearer's beck and call. But he had willingness and aptitude and quickly learnt the knack of picking up. He was soon confidently punching above his weight in the rough-and-ready chiacking of the shearing board. In fact, he was so ever-ready with a lively answer that 'Wide Awake' quickly replaced 'Wait-a-While' as his call sign. Inevitably, this was often shortened to 'Wider'.

The Charleville Show was only a couple of weeks away, and Jimmy Sharman's famous boxing tent would be on hand, challenging the locals to 'step up and take a glove'. I went into training, running, working out on the floor-to-ceiling bag I'd hung on the hut verandah, and offering anyone a workout with the gloves. Often shearing teams could boast a man or two who could use himself, but here I drew a blank. In jest I looked to Jimmy's skinny frame. Wide Awake said he was a lover not a fighter; however, when I finished my skipping routine, he took up the rope and gave a crackerjack exhibition of pepper, left and right cross-overs, doubles and even triples, the like of which you wouldn't see outside a city fighter's gym – or a schoolgirls' playground.

The team watched with renewed interest, but Wide Awake was blown in under two minutes. He folded onto a chair and said, 'I need a gasper to get the lungs back in order.' He lit up while I again studied some faint scarring above and below his left eye and a thickening of the left lobe, which might have been a young cauliflower ear, usually a trademark of boxers.

'Where did you learn to skip?' I asked.

'National service,' Jimmy shot back without blinking. 'They put me in the kitchen because I was too skinny to lug a bloody great rifle and tramp around in army boots. But they made me do a couple of hours a day in the gym to build me up. They soon found out I couldn't lift anything heavier than a packet of smokes, and couldn't lick my little sister in combat – so they put me on skipping and cleaning up the place.'

On Saturday morning I handed Jimmy my spare rifle, an ex-army .303. 'The old digger's mate,' he said, fondling the rifle familiarly. We went pig shooting along a bore drain. Stalking through lignum bushes, Wide Awake suddenly took aim and dropped a galloping boar pig that broke cover 150 yards away before I could fire my scoped .243 Mauser.

'You're faster than Wild Bill Hickok, Wide Awake!'

'Blame it on the Nashos, mate. When I couldn't keep up marching they put me on target practice. I wound up instructor – musta cost the army thousands of quid in ammo.'

Jimmy quickly became the life of the camp; a live-wire popular with all hands. All hands, that is, except The Bald Eagle . . .

In the culinary department The Bald Eagle took over where The Honeyeater left off. They were both gun babblers, conscientious and proud of their cooking and cleanliness. Yarning over the washtubs one Sunday morning, Rooster opined, 'The Eagle is one of the best. I wouldn't say he's superior, but so far he's manufactured a wider menu than The Honeyeater. The only babbler I've seen with a bigger range was old Tivoli Jones. He acquired his handle because he had more variety than the Tivoli Theatre. I reckon I saw old Tivoli bung on thirty different mutton recipes in a week: fried chops, grilled chops, braises, fricassees, casseroles, veal and venison cutlets – the last two was chops prepared in his special marinades. And he had stylish names he used to print fancy and put on a blackboard he carried for the purpose. There was Blue Danube Duck, which was cold mutton fried in savoury dough. A plains turkey became Royal Roast Pheasant. Oyster Soup and Turtle Soup were bloody miracles because they tasted like oysters and turtles, but they was concocted from diced mutton with plenty of onions and herbs and a dash of rum or port wine – and God knows what.'

'And *you* would be a silver-tailed connoisseur of turtle and oyster soup and fine wines?' Les queried.

Rooster grinned. 'Too bloody right, mate! Nothing's too good for the working class!'

'What about his duffs and puddin's, Rooster?'

'Too many to recall, Les.'

I had been earwigging and broke into a traditional shearer's song 'The Station Cook' I had learnt off a record by American folk singer, Burl Ives.

> The song I'm going to sing you will not detain you long;
> It's all about a station cook we had at old Pinyong.
> His pastry was so beautiful, his cooking was so fine
> It gave us all a stomach ache right through the shearing time.

'That'll do, Presser,' Rooster said. 'Never mind belly ache! You're giving me bloody ear ache.'

I grinned. I liked singing and stirring the possum, and I thought it ironic that I should learn the lyrics of old Australian folk songs off an American. Most singers I met in the bush would regale with pop songs from the Hit Parade and old Irish classics. They only knew a few choruses and verses from our great repertoire of bush song with its rebel heart, humour and scorn of privilege.

Unfortunately, along with his culinary skill and dedication, The Eagle soon displayed moodiness and a bullying streak. This drew little comment from his workmates because it was accepted that the better the cook the crankier he was likely to be, although a warning might go around: 'Lookout for the babbler this morning, mate. He's as pissed off as a pregnant wife on washing day.'

Like most bullies The Eagle needed to bolster his self-esteem by brow-beating someone he believed could not retaliate – not someone entirely defenceless, for the bully's ego requires a victim who fights back, but someone who can be antagonised and crushed at the bully's whim, like a cat with a mouse. He picked on Wide Awake because he was an outsider who was lippy enough to make amusement, but not big enough or important enough to strike back effectively.

Running the length of the shearing board eight hours a day, carrying and throwing weighty fleeces, Jimmy burnt up a power of energy and worked up a healthy appetite. Coming up behind him at the mess table one night the big babbler set his target up. 'How many brothers and sisters have you got?' he asked.

'Nine sisters and eight brothers,' Wide Awake jested between bites. The ridiculous reply suited The Eagle's purpose better than the truth, and he swooped. 'You must be the runt of the litter; all runts are gluttons. Leave some for other people, you greedy whelp.' He looked around the table as if expecting an approving laugh, but his caustic attack earned only the censure of silence as Jimmy ignored him, and hungry men continued eating.

The next night when Jimmy scraped his plate and placed it in the wash-up tub, The Eagle grabbed his wrist. 'Can't you do anything right? Scrape it properly, you lazy runt!' Jimmy scraped the plate again, and then said brightly, 'Sorry, *chef*. I'll do better next time.' He went out whistling merrily.

He was a tuneful whistler, and he had a pleasing light tenor voice. At that time blokes were often heard singing and whistling snatches of Hit Parade numbers and old favourites around the huts and on the job, but Jimmy was special: he knew the songs right through and had the confidence to stand up and sing them on request. The Eagle's reaction to Jimmy's whistling was to snarl, 'Shut-up, Runt!

Don't whistle in my kitchen. I'm listening to the radio.' It was an attitude and voice he reserved solely for Jimmy.

Since the success of the shearing industry's epic struggles to unionise against wage slavery in the 1880s and 1890s, the bush workers had harboured a rugged sense of democracy. 'Fair go, mate' had become the first commandment of the itinerant shearer's culture. Thus Les Goodman, as shed rep, had a quiet yarn with Bert and me, his committee men, and Rooster sat in. 'The cook's making it hard for Wide Awake,' the rep declared. 'He's standing over the little bloke. Do you think I should give him a word to go easy?'

'Not on yer Nellie,' Rooster said emphatically. 'Jimmy's cheeky enough to look after himself. Chuck a grenade under The Eagle and London to a brick we lose a good cook. Besides, Wide Awake is giggin' the cook, whistling and stirring him. This morning he told The Bald Eagle he could get him a wig on the cheap, made of canary feathers!'

When the chuckling subsided, Bert said, 'I'm with Rooster. I'm as crooked as the next bloke on standover coves, but let's leave it for a day or two. With a bit of luck the babbler might start picking on someone else – like you, Les, or the classer or the presser.' He added with a chuckle, 'He's too bloody big for me!'

Within a day of our meeting, Jimmy came in to the kitchen wearing thongs. Union men since long before my time had established a basic code of cleanliness, health, dress and respect for each other. The scowling cook singled Jimmy out, ordering him to don shoes. When he came in wearing an unbuttoned cardigan which revealed his puny hairless chest, The Eagle bellowed, 'Go and put a shirt on,

Runt. Have some respect. You're not at home now! Were you brought up in a pigsty?'

I usually finished work about twenty minutes before the boss rang the knock-off bell and this day was no exception. I went back to my room and found that some mail had arrived. Inside a brown foolscap envelope addressed to me was another addressed to Jimmy G in a neat feminine hand. I left it on Jimmy's bed and went for a training run. When I jogged back the shedhand was still in his greasy work clothes – shorts, T-shirt and sandshoes – sitting on his bed, elbows on knees, head in his hands. 'Cheer up, champ,' I said, hurrying out to shower. On my return Jimmy's attitude hadn't changed. 'What's up, mate?' I asked.

Jimmy replied brokenly, 'It's alright. Just my boy's in hospital; broke his thigh. I should be with him.' Although he usually steadfastly refused alcohol while in the bush, Jimmy now joined me in a can of VB.

Leaving him unwashed and hunched in the gloom I went for tea. The Eagle stood over me as I bent to scrape my plate. 'Where's the bloody runt?' he snarled. 'The little shit's trying to annoy me again, is he? Tell him he's got five minutes to get a feed or he can go hungry.'

I picked up a clean plate and loaded it with roast meat and veggies. 'Where do you think you're going with that?' The Eagle bellowed. Ignoring the big man, I took the meal to my mate.

Jimmy ate and showered. He'd often talked about 'little Jim' with pride and humour, but now spoke only to ask if he could borrow a few cans from my carton of VB. 'Go for it, mate,' I said, and he drank beer and smoked and walked till the small hours.

In the morning he drove to the homestead in my VW and phoned Sydney. 'Little Jim is doing alright' was his brief report when he returned, but his mood remained sombre, and all day he went silently about his work, oblivious to the friendly jokes and repartee of his workmates and The Eagle's bullying jibes.

The dining room and kitchen were screened against insects, particularly flies which swarmed in the summer time. With the chill of winter almost on them, however, the insects had vanished and The Eagle had jammed the door wide open to make it easier for him to come and go.

Jimmy followed three or four others into the kitchen and The Eagle swooped. 'Close the door, Runt. Where were you bloody well reared – in a tent or a pigsty?'

Quietly, Jimmy closed the door and served his meal. Absorbed in his personal worries he ate quietly, scraped his plate and left the table.

The Eagle was annoyed by his victim's passivity. He screeched as the rousie approached the exit. 'Don't forget to close the door, Runt! And answer when you're addressed by your betters! Where were you born anyway – in a pigsty or a tent?'

The rep called sternly, 'Lay off the lad, Eagle! He's having a hard enough time; he's got family troubles.'

But The Eagle paid no heed. And neither did Wide Awake, who had turned back from the door. 'Maybe it was a tent,' he snapped. 'Maybe it was *Jimmy Sharman's* tent.'

The cook bore down on him: a trumpeting, charging white elephant, bellowing, 'Lip me in my kitchen, Runt, and I'll kick your arse till it's red raw.'

Wide Awake danced a boxer's side-step, and swung a leaping left-hook from his hip to the butt of The Bald

Eagle's ear. Then he was back in front of his man, dancing, about to fire a finishing combination, but the big man was out on his feet. He sat down with a seventeen-stone thump that shook the kitchen and rattled the crockery.

Half the team rushed to the cook to inspect the damage and raise him to a chair, but Rooster, his contentment fortified by a brace of over-proof Bundy rums, only stopped eating to declare, 'Yer wouldn't read about it in *Pix*! The pygmy burrs up and drops the giant! It's the biggest upset since Jonah swallowed the whale!'

I swiftly followed Jimmy G and found him sitting on his bed, anxious and breathing heavily. 'Can you run me to town, mate?' he asked. 'I guess I better pull out before Frank sacks me.'

'Don't worry about the pitch and toss, mate,' I reassured him, knowing that the boss would be on Jimmy's side. 'The rep and the boss will have a yarn to The Eagle. Everyone knows he had it coming. They just didn't think you had the heft to do it. Les would have pulled him into gear in a day or two.' I broke the seal on a flask of hospital brandy that had been resting in my first-aid kit for half a year. 'Have a shot of this, mate. It'll settle you down. That was the best left-hook I've seen since Punchy Roberts decked me a couple of years ago in Goondiwindi – but, on second thoughts, I didn't see that one coming.'

Wallowing in self-pity, The Eagle retreated to his room. After a few volunteers finished the washing-up and tidied up the kitchen, the team mustered outdoors around the campfire to yarn and laugh over the unscheduled entertainment. A few minutes later, Jimmy – cheered by the flask of brandy – pulled up a stump. He couldn't be induced to expound on his boxing career. 'Just a lucky punch,' he told us. 'Like I said, I'm a lover not a fighter.'

The general opinion was ten to one that The Eagle would roll his swag – until the rep joined them. 'The Eagle has talked things over with me and Frank,' he said, 'and it looks like he's going to stay ... But Wide Awake will have to apologise.'

'Like bloody hell I will!' Jimmy spluttered.

Les laughed and patted the shedhand's shoulder. 'Come on, lad. Consider yourself lucky. The Eagle says if you weren't such a runt he'd have got up and punched you into the middle of next week.' He waited for the laughter to fade before giving me a knowing nod and adding with finality, 'No need to make it a legal matter, Wide Awake. Leave it till the morning and just say you're sorry. Talk it over with your mate.'

Jimmy and I lubricated our souls well into the night over a few tinnies with stories of life and sport – and family.

The atmosphere at breakfast time was as tense as a magistrate's court when Jimmy came in. Fronting The Eagle, he said firmly, 'Sorry about last night, Eagle. I lost my temper.'

'You're lucky I didn't lose my temper,' The Eagle said aggressively. 'We'll settle this after the shed cuts out.' But he didn't look confident; and he failed to meet Jimmy's gaze as they shook hands.

That night after tea the rep and Rooster visited Jimmy and me. 'You did well, Wide Awake,' Les said. 'I reckon you won't have any more trouble with The Eagle. But don't stir him! We don't want to lose a good babbler.'

Rooster added, 'That's right – and look out for the big

bastard in town. If he gets up a skinful of Dutch courage he'll take a swing at you. His size bluffs most blokes, and he picks his mark. And he's a well-known king-hit merchant.'

The team worked on. For the sake of harmony it was taboo to mention the fight in The Eagle's presence, but Jimmy copped plenty of ribbing. Rooster offered to train him for a heavyweight fight with Floyd Patterson, the world champ. 'You'll have to carry a couple of anvils to make up weight, Wider – and give up smoking.' Jimmy said the anvils would be no trouble, but he wouldn't quit the fags.

The Charleville Show came, and I had a couple of scraps in the boxing tent. Despite the urging of his mates to 'Glove up and have a go, Wider,' Jimmy stayed in the bush. I surmised that he was afraid of being recognised.

We cut-out the shed and the team grew from six to eight shearers to sign on at Pinkilla for four weeks' work. The sheep carried a lot of wool, and my maturing young body was suffering by the end of the first week, even as it was strengthened by the rigours of constant heavy labour. I was proud to keep up with the shearers, pressing forty bales a day without working overtime of a night. My technique developed as I adapted to the moods and mechanics of the Ferrier press, and most days during the dinner hour and smokos Jimmy gave a hand lifting butts of bellies and locks, which were bales of oddments from the board that were filled by hand.

By Friday knock-off I was rung-out. My muscles and sinews were drained of all energy save the willpower which drove them. Halfway through the third week, however, my stamina and technique had improved and I was confident

I would complete the contract on my own. But on the Wednesday after smoko the classer braced me. 'Alex seems to think you're working too hard for a young fella. He's putting an offsider on next week, Red Johnson, to give you a blow.'

'What! I'm doing the job, aren't I?' I protested.

'Yes, you're doing a good job. Just settle down and think it over. The fact is the Old Boss wants to keep Red on the payroll. His team cuts out on Friday. There's a bit of crutching for the shearers but no work for the presser. Come July, Alex says there'll be more work than you can poke a stick at – so he wants to hang on to Red.'

My earnings would be cut in half, but it wasn't the loss of money that was the problem: my pride was hurt. 'I understood I had the shed on my own,' I snapped, as I strode angrily into a bin piled with fine wool to gather an armful.

The classer watched me heave armfuls of fleeces into the top box, climb swiftly up and tramp the wool. With both boxes loaded and tramped twice, I mounted the counterweight bag and, driving with my legs off the upright pole, plunged nine feet to the floor, pulling the top box over on its pivot to fit above the bottom box. The two boxes now stood ten feet tall, and the five by three inch greased hardwood spear with its pulleys and wire ropes towered another eight feet. I dropped the spear onto the monkey board on top of the wool, connected the pulley wires to the windlass and took up the slack with the handle, before picking up the iron lever and swinging into a regular stroke to compact 350 pounds of wool into the bottom box.

Sweating freely while recovering my wind with measured breathing, I fastened the bale and rolled it onto the scales, where I weighed and branded it with the station name – Pinkilla – above the wool grading of AAA and the bale number below. Rolling the bale from the scale I felt the usual satisfaction with the sight, weight and firmness of my labour.

The classer had watched patiently for ten minutes. 'I know the Old Boss is rough around the edges,' he said, 'but he's a trump. He looks after his men. He likes you, lad, and he doesn't want to see you bust yourself. You're a lightweight for a presser, you know, and you're still growing.'

'I *know* he's a good boss. My dad often tells me he is and some of you blokes have worked with him for years. But it makes no difference. He didn't say anything about putting another presser on before the sign-on,' I said, selectively forgetting our actual conversation.

'And he didn't say he wouldn't. This is a pretty fast team, and the sheep are cutting more wool than usual. Alex is a man of his word – and proud of it. You'd be well advised not to suggest otherwise.'

'I won't,' I said curtly, before plunging into a bin of combing wool to gather an armful for the next bale. The classer returned to his table, where the waiting fleeces had overflowed onto the floor.

Following my words with the classer I called it a day and hurried to an early shower. I returned to find Jimmy, just arrived from toil, eyeing a bundle of mail the rep had dropped on my bed. I glanced through the addresses and said, 'No luck, Wide Awake.'

The mail came once or twice a week, and most deliveries brought Jimmy a foolscap envelope under my address, containing several letters. Jimmy would read them immediately, his smiles, frowns and chuckles reflecting the content – usually family affairs, some of which he relayed. Little Jimmy, I discovered, was at home and in plaster, but on the mend. I became vaguely familiar with his wife, Janice, and little Jim's kid sister, Carmel, who was 'as sweet as a lolly and bright as a sunflower'. His father and older brother, however, remained obscure. I didn't inquire, but given Jimmy's resume and mail arrangement I wondered if they might be at Her Majesty's pleasure.

Jimmy inquired anxiously, 'What's in the parcel?'

I grinned. 'Mine! Get your eyes off that parcel, Wider! That contains two books I've been waiting for from the Queensland Book Depot.' But when I removed the brown paper, I was disappointed to discover a smaller parcel directed to Jimmy G in a familiar feminine hand. It was Jimmy's turn to grin as he fondled his parcel with delight, then placed it, unopened, in his port and locked the lid.

'A singular and most unexpected reaction, old chap,' I observed in my version of an Oxford accent. 'What the hell's in it? Nitroglycerine or ten quid notes?'

'Just a birthday present, mate. I'll be twenty-six in a couple of weeks. That's when I'll open it.'

I didn't believe him for a second but knew better than to ask. 'Could be nitro, but I'll back a swag of tenners,' I said.

Red Johnson proved to be a jovial big lump of a bloke with a muscular worker's beer gut and a fierce red beard. He was strong alright, but he hadn't developed a rhythmic method of distributing his energy; and as he was making a comeback to wool pressing he wasn't in good shape. Not that it mattered: with two men on the lever for eight shearers we had time to spare.

Although work was a lot easier, I was still resentful of Alex employing a second presser. I was aware the dint to my pride was clouding my mood, but I wanted to build a reputation as a coming gun presser – and having an offsider to keep the wool away for eight shearers didn't fit with this self-image.

The next Monday night Jimmy and I took a shower and drove to Quilpie before tea. We had the usual feed of steak, eggs and chips at a cafe before going to the post office. I phoned Bill Blackwood in Hughenden and booked a pressing run starting the first Monday in July at Aberfoyle station. I then phoned Alex and told him my services wouldn't be available after Pinkilla. The Old Boss didn't hide his hurt, and gave me a memorable dressing-down. Jimmy also gave notice and then rang someone in Sydney and engaged in a brief conversation. 'Janice and the kids left Sydney this morning,' he reported. With all his love of camaraderie Jimmy was cautious and only hinted at his plan. We went to a pub and Jimmy fed the jukebox. Jimmie Rodgers lively hit song 'Honeycomb' came on.

'That's me singing,' Jimmy said, foot tapping and fingers clicking.

'*You?* Bullshit! That's Jimmie Rodgers.'

'No siree,' said Jimmy. 'That's yours truly, Presser – right up front on the vocals. I recorded a few songs when I was a lad.' He sang along with the lyrics, and when the song finished pointed to the name on the label. 'That's me.'

'Fair dinkum! I reckoned it was Jimmie Rodgers himself. You're a bloody wizard, Wider! More tricks than Harry Houdini!'

On the drive back Jimmy revealed that 'the wife's English. She came here as a girl, but her family's gone back to England. She's leaving the car in Grafton – garaged in my cousin's back yard. She and the kids will catch the train to Brisbane; I'll meet her there. You're the only one who knows, Presser. Keep it that way.'

'For sure,' I promised. I reckoned it might be safer to know nothing of Jimmy's doings, but curiosity got the better of me. 'Why the hell are you planting the Holden? I thought you were in the clear with the law.'

'It's not the cops. But I don't trust *them*. They could track Jan's car and pass the word to other people. My cousin is sweet. He'll wait a few months and then bodgie the plates and cash the FJ in Sydney.' I could only guess who 'them' referred to and speculate about the substance of Jimmy's fear. Perhaps it was something to do with the warehouse robbery, and Jimmy, the inside man, had collared more than his share.

'What about that little parcel, Wide Awake?' I queried. 'The one you locked in your port.'

'Just a birthday present, Presser.' He laughed. 'The wife and the kids are flying to England next week. As soon as I can arrange a passport I'll follow.'

The shed cut-out on Thursday afternoon. Jimmy G and I arrived in Charleville in time for a few beers and the mandatory steak, eggs and chips at the Greek's.

Alex had arranged to pay us off at nine o'clock on Friday morning. Jimmy walked around early while I delayed, saying I had a few things to do. In fact, I feared the Old Boss would give me another earful – this time in front of my workmates. He might even give me a *reference*. No doubt they would grin at my embarrassment and give me hell at every chance.

The contractor came down the stairs to meet me. 'You're bloody late,' he said grumpily. I signed for my cheque on the bonnet of my Volkswagen, hoping I was going to get off lightly with a 'Thanks, Boss' and a handshake, but Alex had other ideas.

'I worked you into my top run, son,' he expostulated. 'I kept you in good work on the understanding that you would stick with me in the second half of the year. Now you've pulled out on me just when I need every man I can get. I looked after you, lad, but you're not looking after me. You let me down!'

The chastisement might have been well deserved, however I didn't have time to dwell. The contractor was long in the tooth, but he was also big and active with great bony hands, and had a long reputation as a tough knuckle-man. He advanced aggressively, while I back-pedalled, saying defensively, 'I didn't need any help, Alex! I can bloody well press for eight shearers and stand on my head. You put Red on without even asking me!'

I dodged twice around the Volkswagen, finally off-siding Alex to gain the driver's seat. I found first gear as the Old Boss shouted angry advice through the window. 'You should smarten up, young fella! You're working too hard; and you're too bloody hungry.'

Although I was rattled, I couldn't help laughing as I revved away. The scene must have looked hilarious to the two rouseabouts watching from the upstairs verandah: the old contractor chasing the young wool roller around the Volkswagen. Men would laugh till they doubled in half when the story went the rounds of the watering holes and shearing sheds. Besides, it could have been worse: if I had jobbed the Old Boss I would be marked for hitting an old man – one of the most respected contractors in Queensland – and men who *could* fight would line up to avenge Alex. On the other hand if Alex knocked me leg-up I'd be the butt of pub and shearing-shed jokes for years: 'The *boxing* bloody wool presser! Old Alex knocked him on his arse and busted his eye. Fight? He couldn't beat a carpet with Grandma's broom!'

I thought things over while having a solitary tea and toast at the Greek's. Perhaps I had been edgy from working too hard, and had let false pride get my back up and dictate my course. I began to feel I *had* done the wrong thing by Alex; and I had no doubt my dad, who was a stickler for doing the right thing by everyone and fulfilling obligations, would agree. I never met Alex again, but over the years a nibble of regret would eat at my conscience whenever I yarned with blokes who asked, 'Ever work with Alex Meekin? A champion, the Old Boss, as fair dinkum as sunrise and a heart as big as Phar Lap's.' Looking back, I still wonder if it was shame or a lack of courage that prevented me from apologising.

I drove Jimmy and his worn but elegant leather port to the airport. He was sporting a brand new Canadian lumber jacket with several large pockets. I commented, 'You look mighty spiffy, Wide Awake. Darcy Dugan wears clobber like that when he holds up a bank. No doubt those pockets are just the shot for carrying loot.'

Jimmy smiled enigmatically. 'The port might get lost or stolen, Presser,' he said. 'But I won't! Anyway, mate, thanks for everything. If I think the coast is clear, I'll write.'

He never did. Yet I'm sure Jimmy prospered in his natural habitat, somewhere in a city's concrete canyons. He was a risk taker and a survivor. If Lady Luck gave him a fair throw of the dice, Jimmy would be a winner.

A month or so later I was at the Great Western Hotel in Hughenden with a couple of mates I'd met at Aberfoyle shed: Bob 'The Plumb' Macklin and Doug McMillan, a baby-faced Kiwi shearer. I was having a few cut-out beers with them when I was claimed by Harry – better known as 'Spiv'. I hadn't run across the Spiv since we had worked together as rousies in 1953, but in the way of the shearing fraternity I asked him to join the shout. Harry, who was half-shot, put his money on the bar. 'Gotta tell you this one!' he began enthusiastically. 'I was at this shed near Charleville. And the cook – The Bald Eagle – he starts in on this little rousie. Wide Awake, they call him. The Eagle, who's a big standover bastard, won't lay off; bores it up the rousie all the time! The rousie cops all this shit until one night he leaves the kitchen door open. The Eagle screams at him, "Was yer raised in a tent?"

'The little bloke pipes up: "Yeah – Jimmy Sharman's tent!" And then he hooks him and The Eagle folds up like a lady's fan.'

Young Plumb, who had heard the story from me, looked the Spiv in the eye. '*Really?* I must say a likely story, *old chap*. And you really witnessed this dramatic altercation, *old chap?*'

'Blood oath,' Harry said indignantly. 'On the Bible! Sweetest punch I ever saw. Clean broke The Eagle's jaw in two places, and he broke his leg when he crashed. *I was there.*'

8

BRONCO AND ZULU

I had crossed paths with Doug McMillan around Goondiwindi in early 1958, when I was twenty and Doug was twenty-two. We met again in July 1960, signing on at Aberfoyle in overseer Richie Jack's team of eight shearers. Richie must have been impressed with our work: following Aberfoyle he earmarked Doug and me for his hand-picked team, known as 'guards', and we worked with him in the second half of the year for the next four years. Regular sheds were Eulolo via Julia Creek, employing ten or twelve shearers, and Barenya and Bogunda, south of Hughenden, with eight shearers each. These sheds engaged us for thirteen or fourteen weeks, shearing some 100,000 sheep; other sheds kept us in work till November's end. Appreciating Doug's yarns and laughing banter, barmaids dubbed him the Laughing Kiwi, a soubriquet that stuck and spread.

We soon became best mates. Cashed up after a season's work cut-out in December, we would enjoy a break. We drove to Sydney a few times, but usually looked for good times and amorous adventure along the Gold Coast, where we enjoyed drinking with holidaying workmates and acquaintances from the west, and avoided mobs of high school graduates, the forerunners of 'schoolies'. Doug would shoot through to race meetings or wander the pubs entertaining barmaids while I read and bodysurfed for hours, trying to acquire the skill to ride waves and dodge dumpers. Come nightfall we would check out the floor shows and fair sex at various venues.

Like a lot of Kiwis Doug couldn't swim a stroke and was frightened of sharks. He considered anything deeper than a bath tub dangerous to his snow-white carcass. 'Check the tub for Noah's Arks, mate, before you turn on the shower,' I'd quip, heading for the beach.

'I'll see you tonight, then – if a grey nurse or tiger shark doesn't convert you into shark shit,' he would retort.

At that time all sharks were generally considered to be man-eaters, but I never admitted that I was actually scared stiff, and that swimming out beyond the breakers with a handful of swimmers known colloquially as 'shark-bait' was a terrifying personal challenge I couldn't refuse.

An early riser, I delighted in digging my protesting mate out of bed at sun-up. Rolling him onto the floor usually worked. Strolling along the sand while the rising sun dazzled the tireless Pacific tide we chatted up early-bird sheilas; and now and then we joined a game of beach cricket.

'Courting of the fair sex is the greatest game of all,' Doug said, but he didn't allow romance to interfere with his passion for horse racing and cricket. On holidays Doug took in mid-week and Saturday race meetings, and back in the west he was a star batsman in shearing teams versus locals, from claypan wickets on the Diamantina to matting on concrete at Hebel; and we'd always catch a day at a Gabba test if one was on.

In the sheds he had become a keen reader. Speeding through my copy of Dostoyevski's *Crime and Punishment* in a day, he commented, 'It wouldn't be a bad yarn if he'd get on with it.' He was kinder to Robert Graves' classic *The Greek Myths*. Reckoning the Greek Pantheon's sexy paganism was more his style than the Presbyterianism of his forebears, he would address 'straying sheilas' with, 'Good morning, goddesses! I'll bet you're Athena and you're Aphrodite.' The girls might quicken their pace, but more often stopped to discover if Doug was genuinely flattering or happily nuts.

Doug would fly home for Christmas to the Land of the Long Black Shroud, as it was dubbed because of its reputed dullness, while I would enjoy our usual family Christmas gathering, then laze around reading and relaxing.

One afternoon, after I had made my annual pilgrimage to the magnificent red brick Victorian building housing the Queensland Museum and Art Gallery, I came home for a late lunch with Mum and Dad. My older sister was married by now and my five younger siblings were at work or on school holidays.

For a few days I'd been reluctantly contemplating blistering my hands, hardening my body and reviving my bank account with a couple of weeks of hard work on the pick and shovel offsiding for my uncle Kev who was a drainer with Lew Parkin's plumbing business. As I entered the house I was relieved to hear Dad say, 'Bob Teitzel rang. He said to ring him as soon as you got back.'

Bob Teitzel had all the humour of a hangman's knot and was almost as abrupt. 'I want you at Elmina starting half past seven on Monday. Ten shearers. Given dry weather you'll cut-out in two weeks. I'm starting six teams straight afterwards. Plenty of work right through.'

Zulu and I left Brisbane at seven o'clock on Saturday morning. We took it easy and cruised in to Charleville about 5.30pm, which allowed me plenty of time to enjoy a quiet couple of beers at the Cattle Camp Hotel. The previous year Zulu had enjoyed lapping a rum and milk in the corner of the bar, but he had forfeited this canine privilege and been barred after biting the legs of a couple of brawling boozers. Confused by the ensuing uproar, Zulu ripped the trousers of two urgers, bit me by mistake, and baled up Bronco the barman – who had jumped the bar to stop the blue – in a corner.

I had appealed the sentence, but Bronco maintained a straight face as he declared, 'Zulu is an under-age drinker. I could be pinched for supplying him.' He pinned a typed notice alongside the list of barred violent, alcoholic offenders:

> *Be it known that to wit, one Zulu, a blue and black cattle dog,*
> *is a proven violent offender under the influence*
> *of intoxicating liquor, and is henceforth barred*
> *for life from licensed premises. By order: Bronco Bill.*

The notice was still on the wall but Bronco wasn't in sight. 'Where's Bill?' I asked Marjorie, a middle-aged local woman, as she pulled a beer.

'He pulled the pin last October. Took Millie, that cheeky, carrot-topped housemaid, to see the Melbourne Cup. More fool, he! Only seventeen and a real gold-digger, that one. He had to buy her a flash shiner for her finger before she'd go. Brazen hussy fair bewitched him! I reckon he was silly enough to think he was the first one to pluck her cherry. On her back at fifteen, that one, and now she reckons she's sittin' on a fortune.'

Marjorie placed the beer on the bar. 'She's still away,' she continued, 'but Billy came back flat broke. He was in yesterday, lookin' for a pen. He was barman here for two years, you know. Talk about style! – bow tie an' all – and saved fifteen hundred quid. He was leg-man for a couple of SP bookies. Fair dinkum, he lived the life of Riley and never raised a sweat, and now the silly bugger's going shearing in January. I told him what a fool he was. "I warned you, Billy," I says. "You're not the first to thatch her cottage and buy her furniture." Same old Billy! He laughs and says, "But by crikey, I had a good time! Any bloke who won't spend a thousand smackers on Millie don't deserve to enjoy life." And then he puts a fiver on the slate.'

I left the pub and ate a mixed grill and chips at the Greek's cafe before I drove to a familiar camping spot on the bank of the Warrego River. There I set four lines to catch a yellowbelly or small cod for breakfast, then built a fire and put the billy on and pumped the Aladdin pressure lamp. I was lying on my swag, reading and enjoying solitude beneath the starry dome of the Milky Way when an old ute pulled up twenty yards away. The dog snarled, but I collared him before he could charge.

'I'd know that bloody dog anywhere,' a voice called. 'Don't shoot! It's Boomi and his mad cousin. Tie that bloody dog up before I get out of the car!'

Boomi didn't drink or smoke, but his cousin, Mack, was merrily half-sozzled – his usual state whenever I met him between jobs – and I instantly cottoned on that my plan for a quiet evening was out the window. Mack beheaded a Tooheys longneck with his teeth and handed it to me. He threw tea in the bubbling billy for Boomi.

After we had passed a respectable time yarning and laughing, Boomi broached a subject of deep concern to his cousin. 'Tell the old mate about that roo that ripped your balls out, Mack,' he said, his compassion rivalling his sensitivity.

I had heard a colourful version of the encounter from Boomi in Hughenden the previous year, and when he winked at me I cued, 'Bloody bad luck, Mack. The word on the mulga wire is that an old buck roo punched piss and pick-handles out of you and hopped it with your family jewels.'

Realising that my comment might have been a bit indelicate for the feelings of a young buck who had lost his manhood, I added, 'I was sorry to hear that. The sheilas all along the Gwydir River must be howling buckets of tears.'

'That'd be right!' Boomi confirmed. 'And their dads and husbands unloaded the old twelve bore and put the cattle dog back on the chain when they heard Mack had shot his last bolt.'

Mack was a happy-go-lucky comedian. Wearing the mournful expression that one would expect of a young man who had lost his reason for living, he stood up and ceremoniously dropped his work-worn shearer's dungarees to reveal intact equipment that a jack-donkey might be proud of. 'Have a feel o' this lot and see if they're fair dinkum,' he crowed, and threw his head back to laugh madly at a rising half-moon. 'I gotcha that time, Presser! Well, check out the scars, anyway. But keep yer fingers off John Thomas, the ladies' friend.'

The scars were impressive: the longest ran from the brisket to the left groin. 'Gees, mate, he didn't miss you! Yer lucky only the good die young.'

'My colonial oath, mate! They held me guts in with a wet corn bag till they got me to the Moree Hospital. I was there for three weeks; and then it was three months before I could straddle a saddle.'

Mack beheaded another bottle with his teeth and handed it to me. After casually pulling my swag closer to the car to give himself a back rest, he sat down in one smooth motion, brought his knees up, crossed his dilapidated laughing-side boots and rolled a smoke.

'Why don't you make yourself at home while you're in the Southern Cross Motel,' I said, attempting sarcasm.

Mack opened another longneck with his teeth and then gulped liberally. He then balanced the bottle in his crotch. 'Too right, mate,' he drawled. 'Now I'm comfortable I'll give you an eye-witness account – none of Boomi's bullshit. There was three of us, yer see. We put the nets in first, then made camp and cracked a couple o' bottles. We was havin' a yarn and a smoke when suddenly this big grey buck comes thumping up and squats about five yards away.

'Henry yells, "Holy stuffin' duck shit! Hey! Don't shoot him – he's got a collar on. He must be a station pet!" Then the big bastard hops closer – like he's nearly on top of us – and stands straight up, and growls like a mad dog. I've never seen anything like it. This ole-man bloody roo! Must be seven feet tall on his tip-toes, and he's ready to jump on us. Mick and Henry scramble backwards but I steps up and offers the blighter me bottle. "'Ave a beer, mate," I says. Next thing the ignorant great bastard's got me in a bear hug and is rippin' me guts out with his hoppers. Mick picks up a great tree branch. He yells out, "Duck!" and knocks me cold – and the roo shoots through. I come to and Mick says, "Sorry, mate. Are you okay?"

'"Okay? You bloody clown! How would I be okay? You've spilt me bottle o' beer!" I tell him.'

Mack soon stretched out on my swag, while Boomi and I rested our backs against the VW. The fire faded and we yarned about old mates and past times. Tied to a log, Zulu patiently gnawed a bone and hoped a wild pig would appear, while Mack lifted his head now and then to mention another 'beautiful cocky's daughter' who wanted to marry him, hee-hawed madly, and took a powerful draw on a longneck.

When Boomi rolled out his swag, Mack said, 'Make my bed, too, cobber.' Boomi rolled out Mack's swag, and I said, 'Get off my swag, Mack, or I'll kick yer.'

'Kick away, cobber,' Mack mumbled. I dropped onto his swag and fell asleep.

Boomi woke me about 3am to check the fishing lines. We went to the river and landed two fat yellowbelly. A two-and-a-half pounder and a three pounder, the shearer reckoned. We pulled the lines and packed the fish in mud, then buried them in hot ash and went back to our swags.

Mack splashed his face with river water and greeted sunrise with cooees and yodels before beheading a longneck. He took a long swallow and offered the bottle to me.

'No thank you, Boss, I'd rather not,' I said and sang:

You can talk of your whisky, talk of your beer,
There's something much nicer that's waiting us here,
It sits on the fire beneath the gum tree,
There's nothing much nicer than a billy of tea.

Mack laughed at my rendition of the Australian folk song 'Billy of Tea' and guzzled the last half of his longneck. 'Yer both bloody wowsers,' he declared. 'I reckon I might as well give away trying to educate you blokes about the better things in life.'

I chuckled and improvised:

And a good yellerbelly baked black fella style
Is the best feed you'll find in many a mile.
So tip out your booze and empty your glass
And jam that vile bottle right up your . . . tail pipe.

Boomi spared a grin. 'Another couple of summers and if you don't join AA, Mack, you'll be camped on the creek with the other deros with the arse out of your strides.'

'At least they'll have the manners to share a bottle before breakfast,' Mack quipped, still laughing. 'Not like you ignorant wowsers.'

We breakfasted on baked yellowbelly and campfire toast with billy tea. 'Yer can't beat the old black fella way for cookin' fish,' Mack said. The mud and skin had peeled back and the gut had compressed to a small knot, exposing sweet-smelling delectable white flesh which we sprinkled with salt and pepper.

We packed up and cleaned up the camp. Mack could have had a pen at Elmina, but he had chosen a few weeks' horse-breaking up Augathella way. 'A few busters and kicks don't hurt nearly as much as shearing them bloody great red-eyed wethers,' he declared. 'Boomi loves shearin' so much he's got sheep shit on the brain.'

9

THE JEWEL AND WALLACE

Boomi threw his port and swag into my car. 'I'll take a chance with that biting blue mongrel,' he declared, pointing at Zulu, 'if he stays in the back seat.' We were waiting outside the UNGRA office, yarning with a small muster of shearers and rousies, when Big Bob opened up at eight o'clock to issue riding instructions. Red-faced and unsmiling, he barked orders and road directions, parting with, 'No alcohol on the property.' Experience had taught me that this had become a matter of form: authoritative bluster, unenforceable because good workers wouldn't wear it and reasonable graziers knew it was unfair and stupid. Even the Union thought courts would find it illegal.

I bought a carton of VB tinnies, and we ate an early lunch of steak, eggs and chips at the Greek's cafe then hit the road for Elmina, some eighty miles south of Charleville, via Wyandra.

A few miles from town the bitumen gave way to gravel road, rough with corrugations and potholes. We travelled steadily, hanging on the edge of the dust fog, until we pulled up at the pub in Wyandra. The Sunday session was closing, but we scored a beer for me and a lemonade for Boomi.

'You going to Elmina?' the publican queried. 'Four of your jokers just left. They blued and I sent 'em on their way. You know Bronco Bill? He's alright, but this other bloke – they called him the "Jewel" – is a nasty bit of work. He was looking for a stoush when he came in. I bombed him earlier for swearing in front of the missus and he shut his trap. When I was outside he started again. Bronco chatted to him and got a smack in the kisser.'

'Sounds like Jimmy the Jewel,' I said without affection. Jimmy and I had tangled six months earlier in Hughenden at the Grand Hotel. The Jewel, who was drinking with his mate Leo, had crudely rubbished a young barmaid I was keen on. She was a Sydney nurse on the last leg of hitch-hiking around Australia before heading overseas. I rightly guessed the Jewel had dropped the weights on her and drawn a caustic knock-back, and his jealousy had turned ugly. Instantly angry, I had snarled, 'Outside, you bastard!'

Leo had come between us. Surprised and cowed by the ferocity of my response, the Jewel stuck to his seat as he mumbled, 'I'm too drunk to fight now, *sport*, but I'll loosen you up in the morning.'

'I'll see you tomorrow at eleven o'clock, when you're sober,' I promised and turned for the exit.

'Tomorrow never comes!' the Jewel shouted.

Aware that Leo was a known king-hit artist, I enlisted Torky, a lightweight with a heavyweight punch, to watch my back the next morning. It wasn't difficult: a mad punter, Torky would rather have a blue than back a winner.

My blood was up, and at eleven o'clock sharp an eager Torky followed me through the bat-wing doors into the bar of the Grand Hotel. I had come to knuckle, yet offered a conditional retreat: 'Say out loud, "I'm sorry for what I said last night; what I said was black lies. I was drunk and I apologise." Get that off your chest and then we'll go to the Western so you can tell Faye to her face.'

The Jewel laughed nervously. 'Have a drink, sport,' he said. 'It was only a joke. Like, yer still a bit wet behind the ears – that's yer problem. Yer need to grow up a bit before yer play with the men and the ladies.' Giggling, he turned to his mate. 'What do you reckon, Leo?'

Leo glanced warily at Torky, and said amiably, 'That's right, Presser. Forget about it and have a drink.'

I jerked my man off the barstool in a headlock and hauled him the length of the bar in a cross-buttock before slamming him to the floor. The Jewel's resistance was feeble.

'You want some, too?' Torky snapped hopefully at Leo.

'It's not my blue,' Leo said, and picked up his beer.

'Any old time you want it,' said Torky. The Jewel remained safely seated on the floor. He shouted at me, 'I'll get you! You bloody big-noting standover man. Yer can't even take a joke!'

I exited swiftly, while Torky hung back, disappointed he hadn't been called upon to knuckle.

Six months on, the memory was fresh in my mind as we downed our drinks and hit the road again. Leaving the highway we saw a Holden ute which had bypassed the pub. 'That'll be Yabba's ute,' Boomi declared. 'He takes his time and everyone else's. We'd better slip around him or we won't get to Elmina in time for breakfast.' We went wide on the table-drain and brushed the mulga to get around the dust cloud and the ute.

A flat back tyre lurched our car off the road a few miles on. Flats and blow-outs were par for the course on bush roads, so bush travellers knew their stuff. We got the spare wheel, wheel-brace and jack from under the bonnet, and while I jacked the rear of the V-Dub Boomi worked on the wheel nuts. Yabba's ute slowed to offer help, but we waved him on, and a grinning Yabba and his passenger gave us the finger while the ute accelerated, trailing a billowing red dust cloud which swallowed us and painted the leaves of the roadside mulga trees.

A few miles on we crossed a camel-humped grid and pulled off the road behind a Morris Minor and Yabba's ute. The Minor was jacked up, its wrecked muffler lying beside it. Bronco Bill and the Jewel were trading punches on the road, while Yabba was dancing around, shouting encouragement and refereeing the blue.

'Shut the dog in, Boomi,' I ordered urgently. I was out of the driver's seat in a flash, adrenaline pumping. Bronco went down but got straight up and bored in swinging. The Jewel side-stepped professionally, and dropped his man with a three-punch combination. He scowled down at Bronco. 'Get up! Get up, yer fuckin' smart alec! I'll give yer try to stand over me and call me a foul-mouthed ignorant alky.'

Bronco rolled on to his elbow. His white ironed shirt was smeared ruddy with dust and gore, and blood trickled from his left eye. He spat blood and said, 'If that was yer best punch yer in more strife than Flash Gordon.' He struggled to his feet and staggered for balance. The Jewel jigged in to deliver the finisher, but I jumped in his way and shoved him back with a hand to his chest. 'I'll take his place!' I snapped.

Recognising an old foe, the Jewel snarled, 'Get out of the road. This isn't your fight!'

'It is now! Bronco's had enough,' I declared, shaping up while the Jewel backed off. 'No guts, Jewel! C'mon! Have a go! You weak bastard. You reckoned you were too drunk last time, but you look fit enough now . . . *sport*.'

Standing on the road, surrounded by hostile faces, the Jewel dropped his hands and met my eye in defeat. I felt let down; I had been certain the opportunity to settle the Hughenden confrontation had come. 'C'mon, Jewel. Have a go – or I'll start without you,' I threatened. But the Jewel was deflated: the exhilarating adrenaline surge he'd felt during the fight had evaporated and he walked away.

Having gone through childhood as an undersized asthmatic, I had learnt to box – I believed – to defend against bullying, yet occasionally I caught the breath of the bully in my own actions. There would have been nothing heroic in punishing the Jewel. Such a bout would have been as unequal as the one I had just witnessed. The Jewel was as skilled as me, but I was bigger, fitter, far stronger – and sober. The Jewel knew, and wisely backed off.

A fair-headed young rousie, a passenger in the Morris Minor, grabbed my hand. 'Name's Gordon,' he said, 'but they call me Curly. Pleased to meetcha, mate. Yer did the right thing – only yer should of thumped him! That Jewel bloke is the nastiest bit of work I've seen in a long time. He's been tryin' to pick a blue ever since we left Charleville.'

Yabba, who had done some real fighting against the warriors of Japan in World War II, hadn't seen me since I was a stripling learner wool presser three years prior. Since then I had filled out and matured while learning my trade. Yabba's back was up, but he shook hands before voicing his resentment. 'There was no need for you to come in! I was going to pull 'em up when you put your bib in.'

Carl, a tall, fair, balding young shearer from Wagga gently sponged Bill's battered face with a handkerchief he'd moistened from a waterbag hanging on the side of Yabba's ute. 'I'm marrying on Easter Saturday,' he said, 'and only took a pen shearing Queensland wethers to build up the marriage money. I couldn't fight my way out of a wet paper bag – or I would have left the little shithouse in Wyandra.'

We watched Bronco Bill walk purposefully up to the Jewel, who was standing smoking under a tree some fifty yards away. While Bronco was fair dinkum, fearless and cheerful whatever fate handed out, the Jewel seemed to be forever burdened by a chip on the shoulder. The watchers saw them talk briefly and shake hands before they walked back. The Jewel looked sheepish. Bronco – his eye still oozing blood – laughed as he spoke through split lips, 'She's all over, drover! Jimmy only dropped me five times. He knows he was lucky. Everyone knows Bronco don't warm up till he bites the dust six times.'

Carl said, 'It don't make no difference. He's not riding with me!'

Using fencing pliers he wired the exhaust pipe to the chassis of the Morris Minor with a Cobb and Co hitch.

THE JEWEL AND WALLACE

The load was too heavy for the low-slung car, so Bronco boarded my Volkswagen while the Jewel and his gear transferred from the Minor to the back of Yabba's ute. VW Beetles, with their high road clearance, unique suspension and reliability on bush roads, were marvels, so naturally we took the lead on the rugged road to Elmina, leaving the Minor and Yabba to plug along cautiously in our dust.

After claiming a room and unpacking I went to meet the shed overseer. The staff accommodation was the usual two-bedroom building sheathed in raw galvanised iron lined with masonite, with a small gauzed verandah shaded by box trees.

The dog sat and waited in the shade while I introduced myself to Brian, the overseer-cum-classer, and chatted to Roy, an old acquaintance and the shed 'expert' (the man who sharpened shearers' combs and cutters, and kept the machinery in repair). Brian, a fair-haired slender bloke of about thirty, stopped prepping the books to stand and shake hands. He would prove to be a capable boss, a staff man who followed the book and kept his distance from the men while remaining approachable.

I turned to my dog. 'C'mon, Zulu,' I said. 'We'll say g'day to the babbler.' It was clear Jack wasn't a talkative man, yet he welcomed my hand of friendship. We yarned briefly while I watched him busily preparing tea. He disclosed that he hadn't cooked for shearers for years, but he wanted to get a quid together so he was going to do the full run.

A little above average height, he was straight-backed and not much more than skin and bone. Slow, methodical movement and a furrowed face gave the impression of a man over sixty, but a full head of grey-streaked, wavy black hair made me guess that he might not be over fifty. Spotting Zulu through the gauze door he said, 'If you don't mind I'll give the dog a few tidbits now and then. I get along well with dogs.'

'He won't mind, but if he makes a pest of himself let me know and I'll tie him up.' Zulu wagged his stumpy tail in agreement, ever hopeful a chop would come his way.

As the shed got underway it was tacitly agreed Jack was doing a good if not a brilliant job. Inevitably mutton was the staple provider of protein – roasted, boiled, stewed, curried, fried and grilled – accompanied by baked or boiled tinned veggies. At midday dinner this was followed by milk or rice puddings, tarts, jellies or even occasionally homemade ice cream. Eggs, boiled, scrambled, poached, fried and curried were also served with regularity.

'Another egg and I'll sprout feathers!' a laughing Bronco cracked, and opined, 'Happy Jack don't crack a smile and he don't laugh – like he's the Man in the Iron Mask.' Thereafter the babbler was known as 'Happy Jack'.

The full team of twenty-two men – ten shearers, eight rousies, cook, expert, overseer/classer and me – turned out for tea. John Wallace and I greeted each other gladly. We had met the previous year in Richie Sack 'em Jack's northern run and instantly taken to each other. Wallace was a good-natured, craggy-featured bloke, an inch under six feet, raw-boned, quiet and easygoing. Three years earlier, at the ripe old age of twenty-five, he had taken on the back-breaking task of learning to shear. Now he was getting his 'average' and still improving.

THE JEWEL AND WALLACE

Following breakfast the next morning the team mustered on the hut verandah and elected Yabba shed Union rep and Bronco his committee man. I was appointed second committee man, as this position was by tradition automatically given to the presser. A show of Union tickets was held and the committee checked with the cook to ensure he was satisfied with his set-up, and that the dunnies and the washhouse were in order. A rousie was employed at four bob a man per week to clean the dunnies and washhouse daily and light the coppers for hot showers.

Yabba, as his appellation implied, felt compelled to give a speech. He pointed out that they had all occasionally felt insulted to arrive at quarters which were dirty, below standard and 'lousy with mouse shit and redback spiders'. This place, however, was 'spick and span, and well equipped'. He, and the Union, which had fought hard for better accommodation, expected the members to 'show respect for owners and workmates by keeping it that way, and leaving it the same on cut-out day'. We then headed to the big galvanised-iron shearing shed, where a time-honoured traditional scene unfolded under the scrutiny of the boss-of-the-board and the station owner. The team signed individual contracts and hung waterbags and sweat-towels; the shearers loaded their bog-eyes while the rousies busied themselves sweeping up dust and arranging wool-packs. When we were ready to 'harvest the clip', Brian rang the bell and ten shearers – all 'green' in condition and most short on enthusiasm – tackled 11,000 wethers.

By afternoon smoko a glaring summer sun had converted the big shed into a giant oven. The mercury in my thermometer climbed to 118 degrees Fahrenheit (nearly forty-eight degrees Celsius) at 1.30pm and hung there till 4.30.

On the second afternoon two shearers, carrying some flab and short of working condition, went down with nausea and cramp. Even Boomi, his wiry body unsullied by booze and tobacco, felt short of a gallop and restricted himself to 130 a day those first three days. The heatwave carried on. Still shearing well within himself, Boomi shifted up a cog and sat on 160 on the fourth day. There was no one within a cooee of his tally. The shearers, catching and dragging weighty wethers, bending and driving the bog-eye for eight hours, would drink and sweat a couple of gallons of water a day. The old hands, however, were aware of the dangers of heat stroke and cramp so worked rhythmically and regulated their intake of fluid, salt and sugar. Some swore by quinine and calcium tablets.

The rousies and the classer also streamed sweat and felt the weight of heat oppression, but they weren't bound to heavy labour and the hot bodies of sheep as the shearers were. For me, six weeks of beer and easy living had loaded a stone of fat on my frame, and though I was taking a pinch of salt with a regular swallow of cold sweet tea, cramps and the dry spews grabbed me. As I was a habitual stirrer, I was now punished for my sins. One of my usual capers – when I was ahead of the wool – was to wander along the board and 'stir the possum', as they say. I would pick up a few fleeces and chiack a suffering shearer, saying, 'Gees, mate, you call yourself a Professional Gentleman of the Long Tube; a Bosca of the Bog-eye! You're a bloody presser-starver! Me wife and six billy lids [kids] are going hungry 'cause yer can't undress enough mutton to make an overcoat for Tom Thumb. Avago, mate – ava bloody go! Cripes! I could o' been sitting beneath the palm trees at the Coolangatta pub freezin' me fingers around a pot of Fourex, but I come out here to give you blokes a hand – and what do yer do to me? A schoolgirl could keep the wool away for you snaggers.'

'Piss off, you idiot,' was a common reply. Other responses often implied a low opinion of wool pressers in general and myself in particular. I had been 'barrowing' – learning to shear – in my spare time for years, and could shear my hundred in easy sheep, so now and then a victim of my rough humour would shove a bog-eye into my hand and say, 'You're so smart, Presser! Put yer head below yer arse and avago yerself.' Bronco had his own version: 'Before you piss off, Presser, shear one of these bloody monsters while yer old mate has a smoke.'

At smoko time this particular day, the habitual possum stirrer became victim. As I sat on a four-gallon drum on the shady side of the shed, Boomi came over. I doubled up and heaved. 'What did you say, Presser?' he queried innocently. He turned and called to the team, 'Hard to say what the old mate's on about, but I think he wants more wool.'

'Tell him to keep quiet,' someone called. 'He's making me feel crook.'

When I thought I had finished spewing bile, and was beginning to wonder whether death by strychnine might be preferable, I took some chlorodyne from the shed first-aid kit and mixed fifteen drops of it with water. A few minutes later my stomach began to settle. I dosed myself again on the hour and worked doggedly on.

When the bell ended the last two-hour run of sweat-soaked toil at 5.30pm, Yabba led the way to one of the bountiful wonders of the arid outback: a flowing bore drain. It was bordered for a few feet on either bank by green grass and meandered close by the huts. Easing our weary carcasses into the hot water, we quietly rejoiced at the restorative power of the flowing mineral waters, before emerging after ten minutes to top off with a cold shower. Thereon a soothing dip in the bore drain followed by a cold shower became a regular practice for most of us.

Bronco and Carl placed two planks across the stream to serve as a 'bar' for tinnies and longnecks. These had been covered in hessian earlier in the day and cooled on wet earth in the shade under a 3000-gallon rainwater tank. We would loll and luxuriate in the water, putting the day's toil and aches behind us, and joke about our good luck in finding a holiday resort so far off the beaten track.

One afternoon a dripping Bronco waded ashore in his birthday suit and grabbed an overhead pepperina branch with one hand. Grinning, he yelled, 'Whatever yer do, boys, when yer write home don't tell 'em about this joint. Ain't it heaven on earth: icy-cold beer and hot medicinal waters. If tourists hear about this they'll give Surfers a miss and be here in droves. We'll knock off work and won't get near this lovely beach because those Sydney sheilas in bikinis will swarm all over the place. Yer won't even get strap-hanging room much less a park for the Mercedes or the Rolls. I'm givin' yer the drum: tell 'em nothing and take 'em nowhere.'

When a musical mood struck him Wallace would strum three chords on a beat-up old guitar and sing country songs in a rugged, tuneful baritone. One afternoon, he sat on a four-gallon drum in his dripping-wet underpants and sang a parody of 'The Dying Stockman'. The chorus was well known, and the lads sang lustily:

A strapping young shearer lay dying,
His bog-eye supporting his head;
The rousies around him were crying
And wishing the bastard was dead.

Only Carl's head and a tinnie of VB were above water as he yelled sincerely, 'Good on you, Wallace! You sound a lot like Hank Snow.'

'Good on yer be buggered!' Bronco Bill called. 'More like a squawking bullfrog halfway down the gullet of a black snake.' Wallace let out a wild up-country cooee and belly-flopped into the pool. He ducked Bronco, and bathers cursed and scrambled for shore with their tinnies and long-necks held aloft. 'Bloody brumbies!' and 'Piss off yer mad bastards!' came from all directions as Bronco and Wallace giggled and swore and grappled like schoolboys.

'Grow up, yer mad bastards,' Yabba yelled, struggling to save his beer. Zulu, who always took horse play for an invitation to draw blood, jumped in and joined the melee until I pulled him out by his stumpy tail.

Carl and Wallace usually ran 'dry' in the shed and enjoyed a few drinks in town on weekends, but the heatwave and their lack of shearing condition combined to weaken their resolve. They borrowed a few tinnies on the Tuesday after work, and on Wednesday night decided to drive to Wyandra and pick up a couple of cartons to repay the beer and sustain themselves over the duration. Curly jumped into the rear seat of the Morris, and Carl reluctantly allowed the Jewel to join him. It had been quickly noticed at Elmina that the Jewel and Wallace weren't on speaking terms. The rumour was that the Jewel and Wallace had been workmates a couple of years earlier on, and Wallace and his de facto wife had invited the Jewel to stay at their forty-acre bush block north of the city for a few days. What occurred there was anyone's guess.

I watched the car roar off towards relief in Wyandra, like a perishing camel bolting for an oasis, then went to bed. With the temperature bottoming at about ninety degrees Fahrenheit (thirty-two degrees Celsius), most of us chose to sleep outdoors in the 'Southern Cross Motel' and pulled our stretchers well away from the huts to take advantage of any occasional breezes.

At around 11pm there was a commotion emanating from the gap between the quarters and the kitchen. I arrived with a torch just as Yabba approached wearing leather slippers and a drab dressing gown. Seeing Wallace and the Jewel facing off, Yabba spoke with authority: 'Break it down, you men! Keep it quiet. You know the babbler's got to be up at sparrow's fart to muster breakfast.'

'Sorry, Yabba,' Wallace said. He turned to the Jewel and spoke sharply. 'You hear that? You're drunk. Forget about it.' As he began to walk away the Jewel snarled, 'Don't turn your back on me, you big bastard. You're all bluff! And you're a bloody fool, too – letting a woman come between good mates. That slut you're keeping is having a lend of you. Half a bottle of sherry and she'll open her legs for anyone!'

Wallace wheeled and snapped, 'I'll see you at six o'clock in the morning.'

The Jewel laughed sardonically. 'Tomorrow never comes.'

I had heard it before.

The babbler appeared, carrying a carbide light. 'What's the racket?' he queried sharply. The flickering light dramatised his middle-aged face – a life-map of a spirited, honourable man striving for respect and independence.

Other men emerged from a hot and fitful doze. Most, like me, were wearing only underpants and thongs. 'Sorry, Jack. The boys are out of order,' Yabba said. He added angrily, 'Cor blimey! They should bloody know better – and they bloody well will. I'll call a swarm first thing in the morning and straighten things out. And I'll tuck these ignorant bludgers in their little beds right now.'

The cook said formally, 'I appreciate that, Rep. Good night.'

'What's he whingeing about?' the Jewel demanded loudly. 'He wakes me every night with his bloody nightmares! You must have f—ing well heard him yelling in the middle of the night. Then he's clattering about in the kitchen at cock-crow. Noisy old bastard!'

'Break it down!' Yabba warned. 'Drag your bed outdoors like the rest of us. You sweat all day then you sweat all night. No wonder you can't bloody well sleep. Wake up to yerself!'

'Come on, pack it in,' Carl advised calmly. 'I'm going to hit the sack, and –'

The Jewel truncated Carl's sentence with a straight right to the jaw that dropped the man in his tracks. The big shearer was by choice and inclination a man of peace, but he was also a proud rugby league forward for his home town. Stunned, he rolled onto his feet and automatically went through the motions of playing the ball. Then, bellowing like an enraged bull, he charged and buried his man as a second-rower buries a breakaway half-back. Transformed into a shouting, screaming berserker, he slammed the Jewel's skull up and down on the packed earth until Yabba and I dragged him off. He was still struggling to get free as the Jewel stumbled around the corner, swearing. 'You're all against me. You can all get stuffed.'

We all went back to bed and at 5.30am I was dozing under the sheet I had pulled over my head to keep out the swarming bush flies and the glare of the rising sun, when running feet thumped past my bed. I got a rear view of Wallace rounding the far end of the quarters. The shearer pounded by again. On his third circuit I realised he was warming up for combat, and followed quickly: I wouldn't miss this stoush for quids. Bronco and Yabba – both early risers – were standing smoking under a pepperina tree, where I joined them.

Fist fights were rare in shearing sheds as there was an unwritten rule among Union men and staff that fisticuffs waited until the shed cut-out – by which time the animosity that caused the blue had often perished. However, Yabba realised this case was an exception; that Wallace – a quiet man – was determined to pay his debt.

His mind set, wordless and fixed of eye, Wallace ran past us, jumped the two steps onto the verandah and bored into the Jewel's room. I was right behind him.

'You said tomorrow never comes,' Wallace snapped angrily. 'Well, tomorrow is here! Get up and we'll see if you're as good as you were last night!'

Hungover, the Jewel squirmed and mumbled protests. 'I was drunk. Forget about it. Go away.' He turned to the wall, and Wallace tipped his stretcher violently. The Jewel's bones jarred as he banged on the floor. Wallace hauled him upright, applied a half-nelson to steer him through the door and shoved him stumbling down the steps.

Word of the fight had already spread around the drowsy camp, and most of the team had gathered. Unaware of the night's events, they gazed sleepily in puzzled wonder. Sensing no sympathy among the onlookers, the Jewel backed against the pepperina. He shaped up and snarled, 'You want it, you big standover bastard? Come and get it.'

Yabba shouted, 'It's a fair fight! Give 'em room!' The overseer strode swiftly from his hut as if to intervene, but stopped short and wisely stood back.

Wallace stepped in a pace and feinted. Still trying to shake the hangover from his wits, the Jewel weaved low and fired a combination that Wallace caught on his brawny arms. He retreated a pace, then advanced launching long-armed, hard-knuckled punches with close to thirteen stone of work-hewed muscles behind them. The Jewel ducked low, presenting only his skull and shoulders – an effective defence in a gloved fight with the Marquess of Queensberry presiding, but suicide on the grass. Wallace hammered the soft tissue of his neck and shoulders, then crouched and straightened his man with powerful uppercuts. The sound of his grunts and the thud of knuckles on flesh and bone were amplified by the morning stillness. The Jewel desperately took a side-on guard, but was battered back until he hung in the fork of the tree, stunned and defenceless.

Hovering close Yabba swooped to intervene. 'He's had enough, Wallace,' he shouted. 'That's it!' But Wallace had already stepped back. Breathing deeply he glanced vacantly around the spectators. We stood as if mesmerised as he inspected his bruised and bloody knuckles, and then strode without a word towards the showers.

The Jewel struggled free of the tree fork. Bronco moved to support him, but was rebuffed. 'I'm alright,' the Jewel mumbled as he straightened his shoulders and mounted the steps without glancing at the silent watchers. Bronco and I followed.

'C'mon, mate, we'll get you over to the shower,' Bronco said. The Jewel sat on the one upright bed and coughed blood and mucus into a towel. He breathed painfully, and felt his rib cage. 'The big bastard's broken a couple of ribs,' he said. 'Hand me the rum and the waterbag. The shower can wait.' As we left, he added bitterly, 'Ask Brian to make my tally up and phone for a taxi.'

Regardless of booze or the blues or inner turbulence, the Jewel always cared about his appearance. He had showered and shaved and was wearing an ironed blue shirt and grey slacks when he walked to the shed to get his cheque. He entered through the wool room and shook hands with me. 'No hard feelings, Presser,' he said. 'Tell Bronco the same. As for that other big bludger: tell him he'll keep – and I'll take the slack out of him when I'm fit and sober.'

'Put it down to experience,' I said. 'He's too big for you.'

THE JEWEL AND WALLACE

The taxi brought a replacement shearer before dinner, and returned to Charleville with the Jewel.

Bonded by the mateship of hard yakka and humour the team became work fit and unified, and from then on shearing progressed smoothly to cut-out. On that last day while I cleaned up the wool and the overseer finalised his book-keeping, the vehicles revved away. Carl and the other boys from New South Wales headed south of the border, and Yabba and a couple more homed for Goondiwindi via Charleville. Too impatient to wait for me, Bronco threw his swag into the back of Yabba's ute and jumped in after it. 'Get mobile, Yabba,' he yelled. 'There's a lively filly waiting for the Bronco in the exercise yard. Let's hit the frog and toad.' He sang lustily:

> I've got a girl in Mungindi
> And another one in Goondiwindi;
> And if these sheilas ever meet
> They're bound to kick up a shindi.

I followed them to town in short order and left my swag and dog at a mate's place. Then I headed to the Greek's for steak, egg and chips, and the Cattle Camp pub for a beer and a yarn. I was glad to greet Marjorie at her regular post behind the bar. She was a human adding machine and swift-handed marvel, who could carry five full glasses in one hand while distributing change to drinkers with the other. She was a friend, arbiter and entertainer, and folksy psychologist. Her current 'patient', my mate Wallace, was leaning on the bar, drunk as a lord, repetitiously explaining that he wasn't drowning his sorrows over his lost love – the lady he thumped the Jewel over – he was celebrating his freedom.

Marjorie broke in. 'Closing time! Drink up!' Slopping up beer with a bar towel, she said, 'Friday nights – I dunno why I bother. Shearers big-noting themselves and shearing more sheep in the pub than they have all week on the board – and slopping booze all over the bar.' She glanced accusingly at Wallace before continuing, 'And when I get home at midnight I'm as footsore as Dobbin, and worried about me boys and Mary – they're out God knows where – and the old man is half-shot and thinks he's horny.'

'Cheer up, Mum,' said Wallace, leaning over and puckering his lips. 'Give me a kiss and feel twenty-one again.'

'Close your eyes, lover boy,' she said, and slapped his pouting lips with the soggy beer towel.

'Yer a sloppy kisser, Mum.'

Marjorie laughed and helped me steer him to the V-Dub.

10

COUNTRY GIRLS AND A CRAZY COP

The next night I took a local girl to the movies. She was fun, feisty and teasing, and as lively and lovely as Marilyn Monroe. I called her Marilyn and, as we'd done before, we spent much of our date laughing. Afterwards I drove her home and, at her insistence, parked at her front door, where we had a lusty, giggling smooch until her mother summoned her. Feeling restless after she left, I drove to the river in search of a party. I parked with other vehicles under lofty branches, and walked the thirty yards to the action.

Despite the heat, the ritual big fire had been blazing but had died down to small tongues of flame darting from a bed of coals. Beneath a half-moon three or four couples were kicking up dust to rock'n'roll music issuing from a tape recorder, while about ten blokes and four or five sheilas were standing around, chatting, boozing and laughing.

A half-shot Wallace was there with three off-duty nurses. Two were dancing, the third, a feisty dark-skinned beauty with Asian eyes, was circling the coals with Wallace, the two of them sparring flirtatiously. 'C'mon, Wallace! Walk through the fire and you can have me. C'mon, Wallace, prove your love for me,' she teased him.

The tape ran out and the dancers stopped to watch the fun. One of the lads called, 'Where's your nerve, Wallace? Avago, mate.' The girls whistled, and one of them cried, 'Gees, you're easy, Annette! Make him work for it!' Bronco Bill loomed out of the night behind Wallace. 'Go, Johnny, go!' he urged. 'Holy snappin' duck shit! I'd crawl through hell soaked in kerosene for what Annette's hidin' under her knickers.'

Half-drunk, and enchanted by the temptress on the opposite side of the fire, Wallace stood with a silly grin, as though anchored in cement; then suddenly let go a lusty yodel, kicked off his thongs, and ran through the glowing coals. He grabbed the astonished girl, and the pair folded in to a lusty kiss while the onlookers whistled and cheered wildly.

As Wallace was quickstepping the coals a vague shape loomed up on my right. I took no heed until a weighty open-handed wallop on the cheek sent me stumbling sideways. Expecting further enemy action I skipped a couple of steps for distance before fronting in fighting stance. My assailant was a confident young bloke, above middle height with a clean-cut profile, wearing grey slacks and a summer sports coat. Having got me out of his way, he was intent on watching the lustful embrace.

I took a step forward and was about to challenge him when the newcomer waved a revolver above his head and shouted authoritatively, 'Now hear this! Everyone remain as you are!'

I scuttled back two steps before the gun was waved in my direction. Looking straight up the barrel, I stiffened like a tailor's dummy. Recognition hit me at the sight of the gun: the man was a copper who had been pointed out to me as a complex personality. When he was sober people said that he was a lovely guy, but when he drank he became unhinged. In fact, a year earlier news had spread that the off-duty cop had terrified people by pulling a similar stunt at the same spot. The word was he had fired a couple of shots skywards, but no one had reported the incident to his superiors, perhaps because they regarded the boys in blue as an unassailable unit, with a long arm and a long memory if one of their number was accused or attacked in any way.

My own contact with the man had been indirect. A young part-Aboriginal woman who worked at a pub as a housemaid had gone to the movies with me a few times the previous year. She had surprised me by ending the brief affair with a cool kiss. 'My boyfriend will be back from holidays on Monday,' she said calmly. 'He thinks he owns me. He'll bash me when he hears I've been out with you.'

'Strewth, Lily! Are you fair dinkum?' I had exclaimed in genuine shock. I was angry because she was dropping me cold and wondered why she had come out with me at all if her penalty would be so severe. 'Well, don't tell him,' I spluttered. 'Where the hell's he sprung from, anyway?'

She looked at me as if I were stupid. 'Somebody will tell him; they always do. Don't you know that?' Seeing the look of hurt and puzzlement on my face, she smiled and softened her tone. 'I guess I wanted to go out with you. It's been nice.'

'Well, why the hell don't you leave him?'

'I live in the quarters at the pub,' she said with some impatience. 'He comes in whenever he wants to. That's why I wouldn't allow you in my room.'

'Gees!' I said, sounding a little incredulous. 'If you really want him out, tell him to go. It's your room. If he bails up, see the boss. I don't know him well, but he seems to be a decent bloke. He'll give the bludger the shove.'

'Will he?' she replied sardonically. 'You just don't get it, do you?'

'I get it,' I said sharply. 'Maybe you don't. It's a free country.'

'Sometimes I think you're bloody stupid,' she said. 'He's a cop! And the boss isn't silly enough to get offside with the law. Do I have to spell it out? They wouldn't get off his back. I mean SP betting and after-hours grog . . .'

'Well, your boss is bloody weak!'

'He's not the only one,' she snapped, her dark eyes flashing anger. 'Have a look at me.' She was standing straight, presenting herself, arms down, palms out, her shoulders back, an attractive young woman with pride in her appearance. Her glossy black hair had been waved and styled and a tasteful necklace of imitation pearls contrasted her flawless brown skin.

'Have a good look. Well, what do you see? You haven't got the guts to say it: "She's just another gin." You're no braver than the rest.'

She strode away, and I called, 'But I took you to the pictures. You know I care about you, and I'm not ashamed to be seen with you.' My inept words followed the fading sound of her sandals.

Now, here before me was the man Lily was so scared of. The noise and motion of the party had instantly ceased with the introduction of the gun.

'Stay where you are and nobody will get hurt,' the gunman commanded, breaking the paralysis by waving the weapon above his head.

A dimly lit figure sitting on the outskirts called calmly, 'Put the gun away, Gary, and we'll leave quietly.'

'Shut up, Sandy! Don't speak unless you're addressed,' he warned.

'You're a bloody lunatic,' said a girl under the cover of darkness – and another whispered urgently, 'Shush! Don't stir him.'

Gary (not his real name) waved the gun again then put it in his trouser pocket. He displayed his empty palms, giggled a crazy hyena laugh, and declared, 'Magic! Now you see it now you don't.'

With the gun out of sight people on the edge of the clearing took a chance on flight. They ran as one, while I, slightly behind the gunman's line of sight, took two cautious backward steps before the gunman aimed his weapon at the runners and mouthed loud gunshot noises. They kept running. The people closer to the fire had seized the opportunity to move cautiously. They were on the point of making a dash when the barrel swung back and covered them. 'Halt. I gave you no permission to move.'

Annette called softly, 'Let us go, Gary. We haven't hurt you. I'm your friend.'

'You stay, Annette. You're a slut.'

I was afraid that Wallace would make a reckless charge and relieved to note Annette had a double-handed grip on my mate's right arm.

The runners had reached their cars and were already accelerating away. Still covering the group by the fire, the gunman backed up half a dozen paces, while I slipped away to shelter in the moon shadow cast by a lofty river gum. A dozen paces away Gary stopped and called calmly, 'You can all go, except Annette. Walk, don't run!' He covered those on his left until he realised that Annette and Wallace were quietly escaping to his right. Supporting his right wrist with his left grip, he aimed carefully at them and called, 'I'm a dead shot. Stop or I'll drop you like sitting ducks.'

I was pretty sure the hammer had clicked twice on empty chambers while the gunman was mouthing shots. Assuming the gun hadn't been loaded and the copper's caper was no more than a boozed megalomaniac's grandstanding, I moved quietly to the edge of the shadow, readying for a rush and a flying tackle.

Annette and Wallace kept walking. 'You wouldn't shoot me, Gary,' Annette called, still holding Wallace's right arm in a two-handed vice, forcing him to her offside and shielding him. 'You're my friend. And when you're not drinking you're a good person. I like you, Gary.' She kept walking and talking in a similar vein until the sobering Wallace suddenly shook her grip and they ran for it, his arm about her.

'Run, rabbit, run! Only two bullets left,' Gary called as the gun panned them. They sprinted into the trees and vanished, expecting gunshots, but hearing only a triumphant hyena cackle and a wild shout ringing through the scrub. 'Fools! My pistol isn't loaded. I fooled you, I fooled you.'

A few seconds of silence dominated before I heard Gary snigger, 'Maybe it was loaded?' He gazed upwards and sang boisterously to the moon:

I didn't know the gun was loaded,
And I'm so sorreee, my friend.

Taking aim at the moon, he yelled 'Bang, Bang' as the hammer struck vacant chambers. 'Nope! She sure ain't loaded,' he muttered and dropped the gun in his pocket. Giggling crazily and singing, he swaggered a few steps towards his car before suddenly gazing all around. 'Well, I'll be damned!' he exclaimed, as though amazed at the sudden quiet beauty of night on the river. I waited, then as his head came into range I dropped him with a straight right, tight-knuckled on a steeled arm. I put so much body heft behind the blow that the follow through stumbled me over my victim.

Later, I wondered why I had stayed when I could have retreated. Perhaps it was curiosity, maybe mock heroics – or for Lily. Yet, the punch was hardly planned. I was grinning as I raced for the car, but as I got behind the wheel it hit me that I had 'jobbed a copper'. It was generally accepted that the best you could expect if you hit a copper was arrest on any pretext and a severe bashing.

Wallace and Annette thought I had revved off with the rest until they ran past my car, parked close by with Gary's. They couldn't have been certain Gary's weapon was not loaded, yet they waited to see if I was alright, and then Annette's VW took off just ahead of mine when my lights came on. She braked hard half a mile down the road; we tumbled out and Wallace chuckled, 'You whacked him?' I was hoping the night had covered my folly, and that there were no witnesses. I hesitated and stammered, 'N-n-no. No way in the world, mate. Job a copper! Cripes, I'm not that bloody silly!'

Wallace said, 'We couldn't see anything, but it sounded like a knuckle sandwich to me. And I heard you laugh.'

''Course I laughed. You can't help an Act of God. The mad bastard tripped and fell. And don't you forget it, Wallace. For God's sake, don't put that bullshit around – the bastards would kill me!'

'Settle down,' Wallace said. 'We don't know nothin' about it.'

Annette's compassion and training were roused. 'Alan, did you see if he was okay, or did you just leave him lying there? He could have concussion or a fractured skull if he fell and hit his head on a rock.'

'Like hell! All I did was laugh and leg it!'

'He'll live – worse luck,' was Wallace's heartfelt opinion.

Gary's lights flared, and his car began to turn around. 'Let's hoof it!' Wallace urged.

Annette declared angrily, 'First thing in the morning I'm going to report him; and I want you both to be witnesses. I know the girls will.'

'Sure thing, love,' Wallace said. 'Count me in.' She glanced at me. 'You too, Alan. Ten o'clock at the police station.'

My reply lacked enthusiasm. 'Yeah, I'll be there in the morning, Annette.'

Wallace chuckled. 'Typical woman: wants to save the idiot's life one minute, and the next she wants to put him in jail.'

Annette snapped, 'Into care – not jail. He needs treatment.'

Around the first corner other cars were parked. Bronco and the nurses and a few others were grouped on the side of the road. Annette's Beetle slowed down to allow Wallace to shout, 'Hit the toe – the mad fella is on his way.'

'Never job a cop!' rang in my head. The lights were out at the house and my hosts were asleep. I quietly rolled my swag on the front verandah stretcher I'd been allotted, collected Zulu from the backyard, and left a note: *Change of mind. I'm heading for Brisbane. Thanks for everything, Les and May. See you next week.*

11

A SHEARER WHO WATCHED EAGLES

In the small hours I passed through Wyandra. Feeling like a wanted man, I drove another twenty miles before camping out of sight off the road for a few hours. I arrived at Kahmoo station, a few miles from town, about 10am on Sunday. As the team wouldn't be expected on the property till Tuesday, I reported my presence at the adjacent homestead, where a confident, articulate woman who I assumed to be the boss's wife, greeted me with an amused smile as she quizzed my early arrival. I tightened up, suspecting that the law had already warned her to be on the lookout for a violent criminal, but when she heard my hesitant explanation about wanting a quiet few days out of town to read and write she widened her smile and said, 'There are no stores at the quarters. I'll find you some chops, potatoes, bread – whatever you need.'

A SHEARER WHO WATCHED EAGLES

I thanked her and got permission to light the stove and the kerosene refrigerator, and to hunt pigs on the property. The fresh meat and tucker was a welcome boost to my emergency tucker box: tinned corned beef and peas, baked beans and Sao biscuits, Carnation milk, tea, sugar, rice, condiments and flour.

Certain I was alone in the quarters, I wandered along the verandah, wondering if the sound of my boots – echoing in the hollow rooms – might awaken the psychometric imprints of thousands of hardy toilers; men who had briefly called these huts home, and laughed and yarned their worries away after each day of back-aching, grimy, sweating hardship. No doubt most of them were solid workmates, who were bound by class, cultural birthright and a few fickle gestures of Lady Luck's wand, to a life of manual toil in the eternal quest for the elusive quid.

I was startled from my pondering by a shout from the end room. 'Hey! Who goes there?'

A man was sitting on a stretcher. He licked a Tally-Ho cigarette paper and used it to bookmark a Penguin volume of *The Grapes of Wrath*. He was an inch and a half short of six feet, and his short grey hair and corpulence on a broad frame advertised an age of about forty. He didn't look or speak like your usual bush itinerant battler, and at first I reckoned he might be an overweight, educated chap, perhaps a 'black sheep' banished to the backblocks by a vengeful ex-wife or bad luck with cards and horses.

We shook hands and he introduced himself as Terry. I took a seat on the opposite bunk, and eyed four paperback titles stacked neatly on the bedside table: *Confessions of Felix Krull, Poor Man's Orange, A Farewell to Arms* and *Rights of Man.* Terry looked over the top of his reading specs and said with a measure of self-consciousness, 'I . . . er . . . guess you don't go for this stuff. I picked them up secondhand in Sydney.'

Well pleased, I grinned. 'Would you believe, I actually do read a lot. I like to keep a few on the go at the same time. Just now it's *Memoirs of a Midget*, *The Case Book of Sherlock Holmes* and HG Wells' short stories. All great stuff. I'll tell you this, though: I don't agree with the critics who put Hemingway above Steinbeck. For me Steinbeck has a sort of in-the-belly sense of humour, which puts him up a class – like in *Cannery Row* and *Sweet Thursday*. And he's a great dramatic storyteller. What do you reckon?'

'I haven't read those two, but there's not much to laugh about in this one. It's a good example of fiction being truer than fact: the ruthless hand of fate, the rich administering authority while bullying and robbing the poor, the cruel exploitation of the unlucky; and the empathy that shared hardship breeds. A great story – but how about *For Whom the Bell Tolls*? That's the strongest novel I've read. Perhaps facing death is the core of the story, but it exemplifies the courage of the human spirit in ordinary folk under pressure, regardless of creed or colour.'

I said, 'I've read that empathetic courage is the sustaining principle of humanity. It arises when called upon by shared tragedy and adversity, when pride and greed have fallen. There's any amount of evidence: Hemingway's Spanish war, Jews in the Warsaw Ghetto, prisoners of war in Changi Prison . . .'

'Yeah,' Terry agreed, passively rolling a smoke. 'But some argue faith in God is even more powerful. However, in the trenches and POW prisons our diggers discovered real mateship – and it was based on courage and empathy. But in any situation a few cunning traitors and informers and weak scabs always surface.'

I was impressed by the shearer's knowledge and delivery. My second guess was that Terry might be a member of the Communist Party as a handful of 'comrades' on the shearing boards still believed in outright Marxism. If he was it wouldn't take long for a dobber to report his presence and for his job opportunities to be restricted. The big shearing companies would fit him with the red-ragger stamp, and his 'picture would be turned to the wall', to quote a popular song by Charles Graham. Our Union, a conservative organisation which fought against communism in unions – and members rights, many claimed – would be of little help. To their democratic credit, a couple of office managers for UNGRA and GRAZCOS and many private contractors and graziers had 'fair go, mate' bred in their bones as surely as Tom Brown in his school days; and they would make a principle of employing a productive shearer regardless of his outspoken political persuasion.

Steamy weather had been building up mountains of cloud for days, and early on Monday afternoon a huge storm broke, dropping a four-inch gully-raker in under two hours. The plains were awash, the mustering was abandoned and the sign-on postponed till the following Monday. Terry and I read and yarned and played cards. I found him to be introspective, even sombre; he kept to himself a lot, spoke correctly and rarely swore. He said he was 'giving the grog a break' and taking blood pressure tablets, but he sweated profusely.

When the ground dried out enough to pick a track without bogging ankle deep, I went pig hunting with Zulu and my .243 rifle. The retribution of the law was never far from my mind, and the excitement of running an angry tusker was a hell of a lot better than sitting around the huts waiting for the inevitable approach of a paddy wagon.

A bore-head located in the shearing-shed wool room bubbled 40,000 gallons a day of sparkling hot artesian water. As at Elmina, the bore drain ran by the huts. I stripped to my underpants, submerged slowly in hot water and bathed leisurely. Then I sat to cool and drip dry on a camp stool while enjoying a beer as much as a man destined for a bashing in the lock-up can. I yarned with the dog and watched a pair of wedge-tailed eagles circling each evening against a cloudy multicoloured sky. Terry selected a four-gallon drum for a seat. He cushioned it with a corn bag and sat silently watching the eagles, lost in thought or reading while absorbing the mood of the closing day.

On the Friday evening I was solemnly drinking a bottle when suddenly I broke into chuckles and boisterous laughter. 'Strike a light!' Terry exclaimed. 'No doubt, it's a glorious evening and Wordsworth or Keats might wax poetical, but I doubt they would laugh like a couple of loons. Why the devil are you cackling? Are you drunk?'

'They say the coward dies a thousand deaths, the brave man dies but once,' I said, still giggling.

'That's rubbish! Hemingway had some experience with front-line fire. He says the brave man suffers two thousand deaths if he's intelligent; he just doesn't mention it.'

'That could be so! But I'll crack another bottle and celebrate. I'm free! Free as an eagle.'

Terry fixed me with a solicitous stare. 'You've had something on your mind, cobber. I didn't like to comment, but – like my gran used to say – "You're as nervous as a long-tailed cat in a room full of rocking chairs."'

I was aware I'd been keeping an eye on the road for approaching police cars, but was confident I had masked my apprehension. Apparently, my anxiety had been obvious, but it didn't matter: only minutes ago, while entranced by distant eagles, it had struck me that I had been worrying foolishly. *Be thoughtful* was one of my mantras, so I was surprised to realise that my rationality had been frozen by fear. If the cops really wanted me they wouldn't need Sherlock Holmes to direct their inquiry to Provo Bob's office – and I'd have been in the slammer and copped a biffing days ago.

Maybe Gary never realised that he'd been bushwhacked; that he assumed a drunken stumble and tumble had knocked him cold as a cucumber. And if he had guessed he'd been king-hit he probably wouldn't have a clue who had hit him: most of the hardy young bush workers could and would punch up if called upon, and wouldn't miss a chance to get square on a standover copper who had pulled a gun. Besides, once the girls had reported him to his seniors Gary would have to tread carefully; and on sobering up he would surely have the nous to know that his part in the affair would come out if he falsified a report. The story might well make headlines in the *Sunday Truth* newspaper. If so it would cause a hell of a stir; and it might well finish his career. Time and the Fitzgerald Royal Commission were to prove that Queensland harboured a significant minority of bent and vicious policemen, but they had the cunning to distract from their dark side with propaganda illuminating their heroic dedication. Lowly constables won no awards for foolishly terrorising citizens at barbecues.

I concluded I was in the clear. I pulled a woollen sock over my cold bottle of Bulimba Gold Top Ale to keep it cool, took a long swallow and laughed a week of apprehension away.

'Cripes!' Terry said soberly, 'since you've decided to go troppo, have you got a bottle to spare? Watching wedge-tails spiralling and circling is relaxing, but it builds up a thirst.'

'Help yerself, mate. In the fridge.'

Terry returned with a cold longneck and sent his bottle top flying with a flick of his Log Cabin tobacco tin before sitting down. 'Okay,' he said, 'let us in on the joke?'

'No joke, mate. It's just for a while there I dreamt I was in the Rogue's Gallery – up there with the big guns like Darcy Dugan. Then I woke up – and I can tell you, it's a bloody relief!'

After a thoughtful pause, Terry said, 'You're a bit young to be dodging a wife-and-child maintenance order?'

'Too bloody right I am!' I said, baling up vigorously on the suggestion that I might be one of the irresponsible sires who found refuge in the shearing sheds. But I offered no explanation for my anxiety.

Changing the subject, Terry said, 'I shore in New Zealand a couple of years ago. The sheep are straight and cut like butter. Lovely country: cool, snow on the mountain slopes, and green all over. A nice change, but mate, I reckon for natural beauty you can't beat a Kimberley sunset – although this one goes close. In fact it doesn't matter if you're shearing in the Kimberley or Back-o'-Bourke, sunsets are the richest time of day. They're as old and as glorious as God, yet somehow there's always sadness as well as beauty, like Keats' "Ode To a Nightingale".'

'The only nightingale I've heard was on an old Zonophone record my uncle Gerald has,' I said. 'It was pretty scratchy – maybe the bird didn't get a fair go – but give me a butcher bird's morning song of praise any day.'

'Or the melody of magpies; and the majesty of a wedge-tail measuring the sky,' Terry said.

I was thinking that although the shearing fraternity could boast of mobs of humorists and yarn-spinners and blokes who were wordy about sport and politics, men as sensitive, articulate and widely read as Terry were rare, enriching companions. It was a week of close companionship I would remember. We sat silently, feeling the emerging spirit of night.

A big sand goanna and a small striped grass snake, tongues darting, stalked the small brown frogs which had emerged for the night's hunt for insects along the grassy green banks of the bore drain. Flights of colourful crimson wings, Major Mitchells and quarrions wheeled by; noisy happy jacks ceased their squawk and chatter as they settled in the branches of the big pepperina tree.

Terry pointed to the departing eagles; black specks about to vanish under a darkening brow of cumulus cloud. 'You'll note they fly in pairs, unless they're teaching the young-uns, when it's a family affair.' He looked down wistfully. 'I'll bet they're headed to their home roost on the river. Lucky buggers!' Then he stood up decisively, and sang beneath the starry canopy in a resounding, ringing tenor:

'Mid pleasures and palaces though we may roam,
Be it ever so humble there's no place like home.

The volume and rich tone booming from his great chest astonished me and filled me with sweet melancholy, while it woke the happy jacks to squawking an orchestra of alarm. 'Bloody terrific, mate,' I applauded when he finished the song. 'Where did you learn to sing like that?'

But Terry was lost in reverie and didn't hear the question. 'I sailed on a freighter on the west coast for a few months,' he said, sitting down. 'It was no different from the Kimberley sheds: lonely men got pissed and boasted about their conquests while secretly they longed for a home and their own woman. I used to sit on deck and watch the waves and the sky and recall Homer's *Odyssey*. Three thousand years later the epic still entrances the mind and shakes the heart: a great adventure story with a theme of longing for home and a faithful woman. Again and again the writings of famous men recorded their devotion and inspiration to lovers and wives.' He chuckled and added, 'And it sometimes didn't much matter whose wife it was.'

I said, 'Let's drink to women.'

Terry lifted his longneck. 'I bet you're thinking of that blonde temptress in Charleville. The one you call Marilyn. You should read Rudyard Kipling's *The Female of the Species*, or better still *The Second Sex*. You'd get a clue to what women are really like.'

Grinning, I said, 'Cripes! You sound like a cranky old man.'

'I am – where females are concerned. And I'm only twenty-six, mate – twenty-seven in May.'

'Gees! You look forty! I hope you're not going to tell me the female of the species turned you grey as a badger before you turned twenty-one.'

'No, mate! Q fever sent me grey in a few months. It was diagnosed as malaria at first. It comes back now and then. In fact I've been feeling a bit low these last few days.' He paused. 'Listen, I married a blonde . . . So, you drink to your Marilyn, and I'll toast the eagles – and my little daughter, Jolly. She's named for Eric Jolliffe, the painter and cartoonist.'

'To the eagles, the sunset and your daughter,' I said.

Over the decades I would treasure that evening, and although we worked together on and off over the next few years I never again heard the shy shearer release his soul in song, or speak of his personal feelings.

We took a draw on our longnecks, and after a thoughtful silence Terry stood up and made for the kitchen, saying, 'I'll crack another bottle while you muster some tucker. What's for tea?'

'Grilled chops and boiled spuds, and rice custard and peaches.'

'Gees! The same as the last three days. You're no Tivoli Jones, Presser.'

'And you're a lazy blighter, Terry. It's time you polished your skills in the culinary arts.'

'I never did learn to cook, AJ.'

Fortified by a solid feed and two bottles of beer, I sat outside while Terry cleaned up in the kitchen. Gravely, I addressed the prick-eared dog: 'Trekking the Congo with HM Stanley or searching South America for lost cities is mighty lonely work, Zulu; and talking to a dog and playing euchre with a sad man doesn't compare with "A jug of wine, a loaf of bread – and thou". In the morning, Zulu, we'll take a run to town to phone a blonde sweetie I call Marilyn. And if the wind blows fair we'll set sail for Charleville.'

Left to right: My sister Patty, Mum and me in 1939, when I was fourteen months old. Mum proudly referred to the tent behind us as a marquee. Made of canvas, it was incredibly heavy, so Dad bought a new green Oldsmobile one-ton truck and we went on an adventure to New South Wales.

I took this photograph of Lorraine Crapp and Gary Winram in Townsville, June 1958, when they were training for the Empire Games.

Dinner camp on the way to Lucknow, 1959. Nellie, Zulu and I admire the catch.

Oxton Downs, 1959. Dad and Zulu are front left.

Left to right: My bachelor uncles, Kevin and Gerald, 1959.

Me, mounted to muster sheep, Julia Creek District, 1959.

Giving a Sunday exhibition spar for the boys in Leeson, Winton, 1959. The Toowoomba rousie (right) soon wiped the smile from my face.

From left: Me, Bob Macklin and Doug McMillan, aka the Laughing Kiwi, on the road from Eulolo to Cloncurry, 1960.

Dad and me with Zulu, West End, Brisbane, Christmas 1960.

From left: Spanner Hayes, Mum, Uncle Merv and Mabel (Dad's siblings), West End, Brisbane, 1968.

From left: My three daughters, Michelle, Helen and Jenny, on Michelle's first day of school, 1972.

With Mum and Dad, 1979.

Demonstrating how to shear a sheep in Banjo's Outback Theatre & Woolshed, 2001. I set up the theatre as a tourist attraction in Longreach, Queensland, and ran it successfully for twenty-one years.

12

MARILYN

'Drop me at the doctor's, mate,' Terry said, as he threw a battered leather carry-all onto the back seat. 'After the doc tells me I drink too much I'm going to have a bet or two.' He was grinning widely – unusual for him. Like a lot of toilers after a spell of hard yakka in the bush, he was cheerful at the prospect of the itinerant shearer's day out: a few beers with mates, a few quid on the gee-gees and 'a punt on the pennies'. He might even get lucky with a 'likely starter', an adventurous, itinerant barmaid, or nurse, or a grass widow.

'If I don't see you at the Telegraph Hotel around lunch time, Alan, I'll know you've gone to Charleville chasing a bit of skirt. I'll stay at the pub overnight. I've seen enough eagles and heard enough ghost stories for one week.' Glancing at me, he raised his eyebrows and inquired, 'Are you going to invest in the Bill Waterhouse retirement fund?'

'Not any more, mate. I come from a family of mug punters. I quit the gambling caper a couple of years ago, after I dropped twenty quid on King Darius in the Cup. My uncle Gerald explained that there's thirty-seven mishaps can turn a "sure thing" into an "also ran" between final acceptances and the winning post – and that's if the jockey is fair dinkum.'

'An honest hoop, AJ? You've got to be joking.' He laughed.

I phoned Marilyn at the butcher's shop where she kept the books and worked the till. 'The girls and Wallace reported Gary,' she said. 'No one else fronted – *including you*.' She paused to let the insinuation sink in, and I stammered a lie about having to go to Brisbane.

'Oh yeah! After you dumped poor little me at home and went tomcattin' out to the river. I missed all the fun! Why didn't you take me with you?'

'Gees, darling,' I said, 'fair crack 'o the whip! Your mum would have disqualified me for life.'

Having got her bite, Marilyn giggled. 'Anyway, the story's all over town,' she said. 'And Gary's new moniker is "Billy the Kid". They say he's been sent on sick leave. And Annette reckons you're a gutless wonder. You promised to front at the cop shop but you dingoed.'

Although I knew she was stirring the possum my hackles were starting to rise. 'I can explain – if you'll just dry up for a mo.'

'Oh yeah, and give you time to think up another porky.'

I let that pass. I was thankful I hadn't been mentioned; and the fact that a copper had been king-hit wasn't on the mulga wire meant Wallace and Annette had sealed their lips.

MARILYN

Marilyn continued, 'Anyway, the sergeant listened to what they had to say, and then he said if they signed a false complaint he would haul them into court.'

'Did they sign?'

''Course not! They're not silly. You can't talk, anyway – you're as weak as water.'

Picturing her playful grin as she worked the mechanical till and cradled the phone between the sweet curve of her shoulder and neck, I said, 'How about I hit the track and pick you up at six for dinner and the movies?'

'That's not on the cards, AJ. I know your form – all you want to do is pluck my cherry, and I'm saving that for my husband.'

'Ain't life amazing,' I said in mock falsetto. 'Drongos collect the cherries while handsome young blokes like me wind up with the sour grapes.'

'That's all you deserve. Now, if you were romantic like Johnny Wallace . . .'

'Strike me pink and roan! Wallace romantic? He's about as romantic as Bob Menzies.'

'Johnny is brave and romantic. He walked barefoot over burning coals for love of Annette. She's rapt. She calls him her "Fire-Walker". 'Course, Alan, you could prove your love for me by walking over hot coals – then you could be my sweetheart. Gotta go now . . . I might just be ready if you're at my place at half past six.' She lip-smacked kisses, chuckled amorously and hung up.

The conversation left me smiling, and full of her devilish fun.

Les and May were out when I tied Zulu behind their house, so I didn't have to complicate life with more fibs about a trip to Brisbane. The night at the movies was sweetly frustrating. As usual, Marilyn insisted we park in front of her home; and true to form we had just begun a promising pasho when the verandah lights went on and a voice called – more in command than question: 'Is that you?'

Marilyn and I uncoupled our tongues. 'That's Mum in the doorway,' she said. 'I better go.' She ruffled my hair as she climbed out, then leant in the window to kiss me. 'Goodnight, my love,' she whispered huskily. 'Next time we'll park out by the sale yards.' She gave a honey-sweet, deep chuckle, and I watched her enticing legs run up the stairs.

13

A BALMAIN BOY AND A FIGHTING SCOT

I camped beyond Wyandra, on the same spot I had the previous Sunday, then gave the dog a run and tied him close. It was always wise to secure a four-legged mate. Left loose even the most trusted hound could give way to instinct and take off mustering sheep; and you didn't know where doggers had set dingo traps and laid baits. Although I unrolled my swag on the sundown side of the Beetle, sunrise brought an early invasion of black bush flies. I pulled the sheet over my head and gained another fitful half hour's sleep.

I lazed away the morning and after a feed at a cafe in Cunnamulla, I gave Zulu a pie and browsed around the newsagent's before buying a *Courier Mail*, a *North Queensland Register*, a *Sydney Bulletin*, a *Sporting Life* and a *Wide World* magazine.

Relaxed and at peace with the world, I took a solitary seat in the Telegraph Hotel, sipped a large orange juice and perused the gripping pages of *Wide World*.

The barman chatted as he polished glasses and tidied the bar with a professional touch. 'A good read,' he commented, glancing at the white hunter and the leopard in a colourful life struggle on the front cover. 'But don't believe too much of it. A mate of mine writes a few of those yarns. He wanders about the public library and Taronga Park Zoo, then him and his missus camp within cooee of a Katoomba pub for a week and, would yer believe, he's battled a leopard hand to hand, put the kybosh on a mad witch doctor and walked with zombies.'

The information was a blow to my adventurous heart and my trust in the printed word. But after a few minutes listening I guessed the barman – who introduced himself as Jerry – might be as big a bull artist as his writer mate. He was a chef by trade, he claimed, but only took it on nowadays when he could get a decent contract. After doing construction work for the Snowy Mountains Authority for three years, he had returned to the big smoke to play A-grade rugby league and cricket while earning a crust cooking for pubs and clubs around Sydney, and 'doing a stint' as a radio announcer. 'I'm a Balmain boy at heart,' he declared. 'I was born there, and so were the wife and three kids. Wife's a teacher. We're caravanning around Australia for a year – before the kids go to secondary school.'

Jerry stood a good six feet, was a beefy fifteen stone and looked about mid-thirties. His left ear was slightly cauliflowered, his nose was double dented, and his right cheek wore a long thin scar; yet there was something patrician about his face, and his brown eyes twinkled. He was a talker but he wasn't a skite, and I felt an immediate kinship.

Three drinkers had breasted the rectangular bar on the opposite side of the bar-room while Jerry and I chatted. The newcomers were alert and spoke confidentially. Two were sharp-featured men in their thirties, a bit above average height, skinny and of olive complexion. They wore navy blue suit trousers and white silk shirts, and might pass for city-based commercial travellers. The third man stood out despite his lack of inches: an out-of-fashion brown bowler hat sat confidently atop a bodgie hairstyle, and he was clean-shaven but for long sideburns and a thick ginger moustache. From behind large dark glasses he seemed to have me under scrutiny. I glanced up but didn't recognise him so went on reading.

Other drinkers casually drifted in. One of them was a burly bush worker in worn overalls.

'What'll it be, Fred?' Jerry asked politely.

'Three pots,' Fred said loudly. After skulling the pots of beer, he slapped a ten-pound note on the bar, upended the glass on top of it, and shouted, 'Who was the best man in the bar before I came in?' It was a classic bar-room challenge, rarely seen but universally recognised.

Jerry called, 'Settle down, Fred. Settle down.' Fred banged the bar and again shouted the challenge.

'You're a noisy wee chap,' the man in the bowler hat called in a sharp Scots burr. 'Anything to shut your trap, but make it twenty and it'll be worth my while to take my shirt off.' He put two ten-pound notes on the bar; then dropped his watch and glasses and wallet into his hat and casually told the barman to 'look after this, pal'.

Jerry glared at him. 'Take your brawl outside . . . *pal*. No action inside.'

Cunnamulla and Winton had earned their reputations as the fighting towns of western Queensland. Most pubs boasted a backyard bull-ring, where the young bucks primed by booze would settle arguments.

As Jerry shoved the hat under the bar a middle-aged drinker wearing a suit proffered the Scotsman a friendly smile and a handshake. He introduced himself and said, 'I'd be obliged if you would let it pass. Old Fred here is a client of mine. He gets out of his tree now and then . . . it's only grog talk. No one takes any notice. You and your mates have a drink on me. The old mate is hungover and half-shot. Put your money away. He'll forget about it in a minute.'

Ignoring the handshake, the Scotsman dismissed him with a glinty eye as he pushed past. 'Keep your advice, Jacko, I won't be heeding it. He's a big boy with a big mouth.'

A fight draws spectators like a bush killing block draws flies and meat ants. Smelling blood, the regulars and a dozen drifters pushed through the bat-wing doors, gulped an eye-opener and nodded to Jerry as they left the bar to see the action.

Although only five-and-a-half feet, stripped to his waist the Scot displayed a barrel torso, and flexed the tattooed muscular biceps of a middleweight. He wore grey slacks and shined Julius Marlow shoes. Fred was fighting in overalls; he looked a bit stunned, as though he was surprised he'd found a taker. He was probably on the wrong side of fifty, but was half a foot taller and a couple of stone heavier; and he looked the part as he shaped up in the old-time grass fighter's stance – straight-backed, left fist and leg extended. His challenger, meanwhile, danced and shadow-boxed around the circle of watchers. Suddenly Scotty dropped his arms, laughed derisively and addressed the onlookers. 'Take a gander at him! Who does he think he is? John L Sullivan?'

His quip drew very few chuckles as he swung around to confront the advancing grass fighter.

Fred led with a straight left and right. Displaying a pugilist's deft footwork the Scotsman swayed under and stepped outside, doubled left rips over the heart, and dropped his man with a short right hook to the jaw. The businessman thrust between them as Fred staggered upright. 'That'll do,' he shouted, pushing at the Scotsman – who shouldered him aside before finishing his man with a merciless crushing right. Turning his attention to the businessman, he snapped, 'You'll get yours, too, if you don't shut your trap and get out of my road.'

The businessman hurried for the pub phone to call the law, while the Scotsman did a triumphant jig and clapped his hands as he challenged. 'Has he got any mates want to try their luck with wee Scotty?' While most silently glared disgust at this arrogant grandstanding, one scrawny old timer shouted belligerently, 'Hang around for five minutes, sport, and I'll fetch half a dozen boys who'll tan yer bloody hide for peanuts, yer loud-mouth bastard. Pickin' on poor old Fred! Yer nothin' but a pie-eater!'

'That's the spirit, Grandpa,' Scotty patronised, and fixed on me. 'You remember me, don't you?' he accused. 'Yes! You! The bloke with the bugle big enough to pipe the Last Post.'

I had long before learnt to laugh with mates making fun of my prominent nose, but strangers and acquaintances taking liberties got my back up. I had recognised the Scotsman, sans glasses and hat, as he stripped; pictured him clean-shaven, crew-cut and vicious, as he had appeared at Malboona shed during our violent encounter three years earlier.

His annihilation of the ageing, drink-sodden Fred was ruthlessly efficient, but any sober bush pug could put together a combination of punches which would KO the likes of Fred. It would be a hard fight, and there was no way out of it with self-respect intact, but I was confident I could take him in a clean fight – or under Rafferty's rules. Scotty remembered me, but he'd be shocked by the strength, experience and confidence his opponent had gained in three years.

Even so, I didn't like the odds: there wasn't a likely looking referee among the bystanders, I had no back stop, and I knew from experience that Scotty would respect the Marquess of Queensberry only while he was winning – if at all. Besides, the Scotsman's hatchet-faced offsiders looked like they would pawn their grandmother's false gold tooth, or gang-bash a schoolboy for his lunch money.

Scotty was grandstanding again, fancy dancing about the centre of the ring of spectators, laughing and addressing them. It was the display of a stoush artist who gloried in picking a blue – and didn't doubt the outcome. He played the audience, revelling in their attention, as much to bolster his own confidence as to destroy his opponent's. Grinning, he jigged and jibed. 'The laddie will show the white feather to the wee Scot. He's a gutless wonder; his lovely mother suckled a squib.'

I felt my gut harden as insults turned adrenaline into anger and fuelled hunger for revenge. I kicked off my thongs to fight in shorts and T-shirt; I'd seen mugs king-hit as they struggled to remove a shirt. Besides, the fabric would absorb punches a trifle and hide the damage of body blows. As I shaped up and manoeuvred for an opening, I was relieved by Terry's unmistakable voice booming support, conveying both promise and threat. 'I'm with you, mate. Get on with the job. I'll keep these toe-rags in their place.' The big shearer had come in late, summed up the situation, and taken a stand close behind Scotty's flunkies.

I struck first with a straight left, which crunched into the top of Scotty's ducking skull, jarring my knuckles. Then he was in, ripping punches into my ribs that had me gasping. He side-stepped to swing a 'finisher' to the jaw, but I got a forearm around his neck, jerked his head downwards and dug in a couple of counter punches. The Scotsman had the best of the exchange. He snapped, 'Fight fair, you bastard. No wrestling,' as I skipped out of range to gather my wits and my wind.

The watchers were quietly intent, while the fighters grunted and sucked breath. I was too engrossed in the contest to heed the thud of fist on flesh and bone, and abusive chat from Scotty. I ducked or parried punches, and from long-range I feinted for the head and punched for the body, while Scotty's tactic was to duck and weave and close, so his pumping short arms could dictate the mill.

I was a stone heavier and in better condition. Sensing Scotty was running out of steam, I went in hard. Our skulls clashed, and then my face was buried in his sweaty right shoulder, and I was pile-driving punches into his rib cage and belly. Driven backwards into the scrambling spectators, who were yelling for his blood, Scotty dropped, winded. There was no referee and no count. Cannily, he took his time while I waited and jibed, 'C'mon! Yer counted out. Yer lovely mother raised a squib.'

Scotty got up, pulling wind and clearly weakened. Over confident, I reckoned he was ready for the 'finisher', but he ducked my knockout punch, stepped swiftly and speared me head-first into the grass with a hip throw. Down and desperate, I parried his following kick with a forearm and brought him down with a leg throw. Now it was a berserker's brawl. I was on top, punching furiously while one of the toe-rags was kicking my ribs – until Terry nearly beheaded him with a back-handed chop. Jerry pulled me off, shouting, 'Break it up. Move! The cops will be here in a minute.'

Scotty was breathing hard, bleeding and spitting chips. A born warrior, he snarled at me, 'If you want some more, follow me!'

My blood was up, and I moved to follow, but Terry grabbed my arm. 'Steady on. Wait a bit! They've had enough – give 'em time to leg it.'

Jerry handed me a double brandy and soda. 'Get this in to you, cobber,' he commanded, 'and then slip upstairs to room five. Clean up, and stay there till I give you the all-clear.'

I took stock in the mirror in room five. For all the ferocity of the battle, I had come out well: eyes and nose unmarked, a split lip, a bruised right ear trickling some claret, and a few hurtful bumps about the rib cage.

I joined Terry, and Jerry pulled me a beer and leant confidentially on the bar. 'A couple of the Queen's finest called. They had a drink on the house and departed. But I'll give you a piece of advice, son: keep out of the Scotty's way. He's a bad-un!'

I was still riding high. 'Be buggered!' I protested. 'I had the bastard beat. He was blowing like a broken-winded nag. I'll see if I can line up a hundred quid bet and a fair dinkum ref, and I'll punch holes in him.'

Jerry spoke seriously. 'My guess is they've hit the toe already. They're Melbourne boys – wharfies' mob. The law said they've got form. Apparently Scotty isn't long out, having done time for armed robbery. You did well. He's fought prelims and a ten-rounder or two at West Melbourne. Could be you're lucky he isn't in shape.'

The news flattered me. I laughed and said, 'Strewth! The way he started I reckoned he was Ray Robinson in Madison Square Garden. Anyway, Jerry, how did you get this info?'

Jerry grinned and winked. 'You give a bit, you get a bit.'

Terry asked quietly, 'What are they doing out this way?'

'Lying low, I guess. They've been at a shed for three weeks. Scotty was shearing and his mates were shedhands. They cut-out on Friday and came to town last night and got on the piss. After this turnout they'll shoot through.'

I laughed. 'Lying low? Crikey! They stuck out like a back-to-front collar in a brothel. When they came in I thought it was Darcy Dugan's gang casing the joint.' The fight had focused all my attention; now I noticed that Terry was wearing a beige summer sports coat and green tie, a rare turnout for a shearer's Sunday drinking session.

'Holy dooley! What's the occasion?' I queried.

The staunch shearer mumbled to the bar top, 'Church... I go to church now and then.'

Returning, the barman refilled our glasses with an orange juice for Terry and a beer for me.

'I'm closing up at one sharp,' he declared. 'Pub cricket comp. I'm the skipper. Would you blokes like a hit? Do you good, Presser – to run some of the bruises out.'

'Suits me,' I said. 'You too, Terry! You told me you cracked a ton or two for the old school, and you look as if a run'll do yer good – if it doesn't kill yer.'

The barman went on his round, and Terry spoke to me confidentially. 'Keep it quiet, mate, but the doc said I've got a dose of clap. No grog while I'm on penicillin. He gave me an injection. I need two more to clean it up. Would you do me the favour of running me to town on Monday night and Wednesday night?'

'No trouble, Terry,' I said, but I was flabbergasted. Since penicillin had become available, gonorrhoea had become rare in the outback towns, yet the mythology persisted: *'You can catch it off a dunny seat, mate, or kissing a sheila whose got a load; and a cook whose got a dose will turn fresh meat bad in an hour.'*

I knew it was bullshit, yet I caught myself shrinking back from Terry. Terry caught on and chuckled sardonically. 'It's not the Black Arabian Pox,' he said quietly.

Later, he said privately, 'Talk about stiff luck: the last woman I cohabitated with was my wife, and that was three years ago. I landed in Sydney and felt like I needed a woman – for the first time in years. I never had much courage to go dancing, so I slung a taxi driver a fiver. He introduced me to this attractive young blonde. Moira was stylishly turned out, she had manners, was reserved and well spoken – you'd reckon she graduated from a posh young ladies finishing college.'

'Maybe she did,' I said.

'More likely the University of Hard Cocks,' Terry said shortly.

'Only a fiver, and a dose chucked in for a night at her place?' I queried.

'Yeah, plus another forty quid, and dinner and drinks.'

The sun was a scorcher when we took the field – about seventeen players a side. Jerry won the toss and padded up with Terry. The medium-paced attack wouldn't attract the state selectors, but it was good enough to allow the paunchy shearer to demonstrate that his natural ball sense had been enhanced by public school coaching. Penetrating the cluttered field, he machine-gunned the boundaries with drives, cuts and hooks, and occasionally lumbered between wickets with the stately reluctance of an ageing WG Grace. He retired on thirty-six, blowing like a beached whale.

At the other end Jerry, opting for over the top, skied out on fifty. He declared on 120, and then took six wickets in three eight-ball overs. The opposing top order survived rockets which hummed about their ears as they went through to the distant keeper, before being bowled by full tosses they mistook for hand grenades while defending from square leg. They were not happy.

Their skipper bailed Jerry up. 'You're new here, big fella. But you ain't Keith Miller and we're not playing for the Ashes! These blokes have got to be at work on Monday. The doctor's not on duty to patch 'em up, and we ain't got a first-aid kit.' Stabbing an index finger into the big bowler's wish bone, he roared, 'Pub rules: nobody bowls more than three overs. You're off.'

The opposition mustered sixteen out for eighty before both teams adjourned to the pub for a free round and a loser's shout.

14

HAPPY JACK'S LAST POST

The owners wanted ten shearers at Kahmoo, some eleven miles from Cunnamulla, but only eight could be found: Yabba, Carl, Terry, Scrubber, a couple of young cocky's sons from New South Wales, and Les and Frankie (two boozy itinerants who said they hailed from 'not far from a pub'). Brian was classer/overseer, Happy Jack was the babbler, Curly and a Cunnamulla lad were picker-ups, and three middle-aged blokes from Charleville were wool rollers and piece-pickers.

I wasn't overjoyed to see Scrubber, recalling a sawn-off scrapper who'd tried to punch my lights out in 'friendly' spars three years earlier. A Goondiwindi lad, I met him when he arrived at a district shed in a VW Beetle he'd paid a deposit on six months earlier. It seemed he took 'Drive away, no more to pay' literally, for he boasted he hadn't made a single repayment since.

We had been tucking into dinner a few days later when two young blokes drove up in a Holden sedan. 'Mizzenmast' Fred, our boisterous Swedish babbler, invited Callum and Noel to take a plate, as was the custom. Mizzenmast, who had sailed the seven seas before jumping ship in Australia in 1934, amused us with stories about the wild-looking Nordic gods and shameless naked women tattooed on his arms and back.

When the visitors innocently admitted they were repossession agents, Mizzenmast exploded into profanity in several languages and banished them to a table on the verandah. 'Out, out,' he raged. 'You are capitalist crawlers, not fit to eat vit good men.'

We were silently slipping into plum duff with custard when a rousie yelled, 'Hey! Those jokers are pinchin' Scrubber's car!'

Men boiled from the mess, screaming abuse and curses, and war-danced around the V-Dub. Mizzenmast wielded a meat cleaver which went close to accidentally beheading Scrubber as he took a punch at the callow youth behind the wheel. But the doors were locked and the windows wound, and Scrubber's knuckles crunched on the shatterproof glass. He swore in frustration as the terrified driver revved the motor. The Beetle might have bowled over half the team, but the other half had lifted the rear end and suspended it on two four-gallon drums; the motor roared and the wheels spun madly, but there was no take-off, and the boys dropped back with jeers and cheers until the engine stopped.

Yabba, as usual, was the Union rep. After relieving Mizzenmast of the meat-axe, the ex-platoon sergeant left the Volkswagen to his troops and besieged headquarters. He swung the door of the Holden sedan open, ordered the repossession agent out, and bellowed at the jostling team: 'Righto! Give us room and keep quiet, you blokes. Let this man have his say and we'll make some sense out of this.'

You had to admire Callum's coolness under fire. He eased from behind the wheel, lit a man-sized Marlboro and passed the pack around. Apart from Scrubber's threats, the team held its peace while negotiations went on for a few minutes. Arthur, the amiable overseer/classer, was six foot two, fifteen stone and nineteen years of age. He said Scrubber had enough sheep shorn to pay for two monthly instalments. Scrubber was an argumentative little bugger at the best of times and he didn't like it, but Yabba twisted his ear and roared, 'You're copping it sweet, boy. Now sign this cheque over and shut up.'

Scrubber paid two instalments and kept the VW – on the condition that he would come good with the arrears within six months. A couple of years later at Aberfoyle shearing shed via Hughenden, when Mizzenmast, Scrubber and I next worked together, the babbler jocularly remarked that the battered Beetle looked as if it had been parked on an army mortar range.

'Oh yeah!' the prickly little shearer perked. 'And I haven't paid those smart bastards another brass razoo.'

We signed on at Kahmoo on a steamy Monday morning. The team pulled into gear, and I wandered about the shearing shed, waiting for the bins to gather enough wool to bale up.

From a cocky's or contract presser's point of view the team didn't look too bright. The brothers were only advanced learner shearers, and Scrubber and Les had hangovers. They were all clearly snaggers, using jabs and short blows rather than long rhythmic movements. Frankie was also boozed out, but he patterned his sheep with stylish easy rhythm, and clearly he had been a good 'un – maybe even a gun – before grog took its toll. Terry, too, had the inherent rhythm of the born athlete, but the fat man was distressed, hanging over the pen gate after every few sheep. Dripping sweat, he said, 'You and your bloody cricket, Alan. I'm as stiff and sore as a drover's dog.' I shore a few ewes to give him a spell.

Yabba and Carl were only solid, average shearers, but they had job pride and were work fit, and they took the lead shearing dense-wooled merino ewes – not large or 'necky', but hard pushing. If the station owners reasonably expected eight shearers to tally 1000 shornies plus per day they were disappointed with 700 per day that first week; especially as they had wanted a full board of ten shearers.

On Tuesday and Wednesday nights Scrubber and Les and Frankie piled into Scrubber's VW and hit the pub till late. Grog sick again on the following mornings, their work did little to boost the team's tallies. The overseer was disappointed, but he couldn't do much about it; he couldn't sack men for boozing in their own time as long as they were on the job and shore cleanly; and with the shearing season well underway reliable shearers were as hard to find as feathered frogs.

By Friday Terry had finished his course of antibiotics, and four days of shearing in a furnace had sweated a stone off him. He punished himself by moving up a couple of gears, and Frankie, who had given his metabolism a break by hitting the sack early on Thursday night, went with him. After smoko, Brian commented sarcastically to me, 'Seven hundred and fifty counted out and a run to go. I've got a team of guns!'

That evening, I grinned and said privately to an exhausted Terry, 'A hundred and fifty's not a bad tally for a fat man with a broken heart, recovering from a dose of the clap.'

'Thank Christ I've finished the penicillin!' Terry declared. 'Hand me a longneck, mate.'

I had expected to have ten solid shearers on the board, hard yakka for at least forty-odd bales a day, and a fat cheque at cut-out. 'Set the pace, Terry,' I said, 'and I might still earn a quid in this cracker.' But that first week wasn't promising: half the team it seemed were alcoholics, more interested in complaining than yarning or cards. After tea I drove to Cunnamulla with Terry and Carl and Curly for a few cheerful revivers. We were back by ten o'clock, singing and joking, a bad week behind us.

On Saturday morning Carl and Curly and I shook off mild hangovers and joined the brothers fossicking through the station rubbish dump. For hours we inspected the discarded power sources and transports of earlier generations: wrecks of old steam and fuel stationary engines, a wool-wagon, a buggy, sulkies, trucks and cars. Before the decade was out city collectors would discover these unattended outback museums, and carry away for restoration countless relics of horse-drawn society and hundreds of decaying Chevs and Whippets, T-Model Fords, Willys and Hupmobiles, and occasionally the corpse of a Mercedes,

Itala or Packard that had once been the carriage of a proud western wool baron.

Earlier visitors had 'bower-birded' the lighter brass and copper fittings. Now the brothers smashed old batteries and extracted the lead and zinc content for sale as scrap metal, and we loaded their booty aboard their Austin Champ four-wheel drive.

On Sunday morning Zulu let it be known he expected to go pig hunting, so we set off with Carl, Curly and the brothers. We followed a creek until Zulu, scouting ahead, put up a big black-and-white spotted boar. While the battle-scarred tusker sprinted towards the cover of dense lignum scrub the dog closed and 'lugged' (seized the pig's ear), and 250 pounds of wild boar and forty pounds of furious dog disappeared into the lignum, making a fearful hullabaloo of squeals, grunts and growls. The boar charged along the pig tunnels, swinging his huge tusked head from side to side and viciously thrashing the dog against the wiry branches and stalks of lignum bush to throw him. With the blood of battle in his nostrils, however, only disablement or death could dislodge Zulu. The shoulder-high lignum tore at my clothes and skin as I followed the sound of struggle, sometimes dropping to all fours and scrambling along the pig tunnels. The brothers followed me, but the men from Walgett – wild boar hunters from way back – sprinted around the perimeter of the thicket and met the emerging life-and-death struggle head on as the combatants burst into the open.

I crashed out of the lignum as Curly dashed in and seized a back leg. Carl threw the boar with a front-leg heave and pinned him with his knee, swiftly unsheathed his hunting knife and thrust the blade into the heart as hunters have done since stone-age men fixed flint to wooden shafts. The great beast let out a dreadful death squeal as bubbling blood gushed out of him and his life force was extinguished in seconds.

Carl cuffed the dog off and wiped gore from his blade on the coarse hair of the boar. 'That'll do, boy. You did well.'

Zulu flopped, panting beside his kill, and lapped bright-red blood. The rank smell of wild boar dominated the sweaty stench of men and the mingling aromas of the bush.

Yabba stayed in the huts over the weekend, while the wool rollers joined Scrubber, Frankie and Les in town, pub-crawling, punting with the SP bookies, and playing poker. They returned to the huts only occasionally to sleep and eat.

After stumbling through a heavy weekend, topped off with a return to the pub for the Sunday arvo session, Frankie, Les and Scrubber were as enthusiastic as Aussie batsmen facing bodyline when the bell rang on Monday morning. They shore along with the learners on about twenty-three a run, while Terry tallied forty and Carl and Yabba 'undressed' thirty-three.

The going got tougher when they cut-out the last of the ewes, and began shearing the two-tooth (a sheep's first adult teeth) wethers. 'They're tough little blighters,' Yabba said at smoko, as he yabbered on about the varying cutting qualities of the different mobs of sheep he had shorn from Tassie to north-west Queensland. 'I've shore plum jams [lambs] and stud rams at Wollogorang and Willandra in New South Wales; signed on at Mt Poole in the Corner Country and Mt Marlow on the Barcoo. Eulolo out of

Julia Creek was good sheep, but they're not all sweet up north. Dagworth, west of Winton, was a tough cut.'

Having been ear-bashed by Yabba's repetitive personal shearing history for weeks, I interrupted flippantly, 'I reckon you must have been rep at Dagworth in 1895, Yabba, and given Banjo a hand to write "Waltzing Matilda".'

Yabba didn't miss a beat. 'Too correct, mate! And I was the picker-up when Jackie Howe tallied three hundred and twenty-one at Alice Downs in 1892. There were a hundred shearers on the board, and Big Jack was on the last stand, two hundred yards from the wool tables. I was pickin' up two thousand fleeces and wearin' out a pair o' boots every day. The pitch-n-toss [boss] patrolled the board in a horse and sulky, and the babbler paddled around the stew in a canoe to stir it. Now, as I was sayin' before you interrupted . . .'

Although we had showered, we workers were still sweating when we gathered along the big mess table that night an hour and a half after work ceased. We ate a bowl of nutritious mutton and veggie soup, and were busy on the main course of grilled chops and salad, or chops and gravy with mashed spuds and boiled pumpkin and cabbage, when Frankie advised slyly, 'Look out for the chops, boys – they're off.'

The advice put the brakes on appetites just warming to the task. Most ceased chewing and used fingers or forks to sniff the suspect meat, while Les shoved his plate aside. 'Yeah, yer right, Frankie,' he agreed. More plates were pushed aside before Scrubber woke up to what was going on. He spat a gob of half-masticated mutton on to his plate, shoved it to the middle of the table and stood up. 'The meat is rotten and this shed is up to shit,' he proclaimed. The vehemence of his outburst surprised everyone. He stood alone, dumbly looking around for support. Yabba hadn't stopped chewing. He swallowed, before bellowing in his sergeant's voice, 'Sit down, Scrubber! Quick and lively!' Scrubber dropped, and Yabba declared, 'Nothing wrong with my chops. Sweet as a nut.'

Curly gingerly recovered a mouthful of chewed mutton with his fingers and placed it on the edge of his plate. 'I reckon it's a bit off!' Curly warned.

Yabba reached for the remainder of the lad's loin chop and gnawed the meat off the bone. 'Yer dunno what yer talkin' about, boy,' he said gruffly. 'This meat is as sweet as mother's milk.'

I had resumed eating. 'You're right, Yabba,' I said, enunciating deliberately. 'I don't know about anyone else's mutton, but mine's okay. Bloody good juicy loin chops, if you ask me – just needs a bit of black horse and salt.' I reached for the mandatory Holbrooks sauce bottle and the Cerebos salt shaker.

Happy Jack heard the commotion. He stood listening for a few seconds, and then exited without a word as Frankie declared officiously, 'This is not good enough, Rep. I won't wear stinking meat. We'll have to have a meeting, quick smart.'

The overseer and the expert were dining in the staff room, separated from the 'men'. They weren't AWU members, so were excluded from the meeting, while cooks usually didn't attend meetings that concerned them personally.

The team assembled on the hut verandah, sitting on stretchers which had been moved out of the sweltering rooms. Yabba sat at a small table and opened the meeting in his official voice. 'As a member of the Union the babbler is entitled to attend this meeting. I've just had a word with him, and he says he has work to do. I hereby extend Jack's apology.'

Frankie said, 'Well, let's get down to it. I want to go to bed. The fact is the chops were off. It means the cook is not doing his job.'

Les was a born 'pointer' and whinger – small in character and low on courage. He was a back-stabber who wouldn't speak out unless he had a charge of Dutch courage and plenty of back-up. 'Yer right, Frankie,' he said. 'I'm not going to pay good money for a man who's not doin' his job. Blind Freddie knows it's no good bombin' a cook. Bomb a cook and they'll get square – like piss in the tea urn or put Epsom salts in the curry. How would we know? I say we spear him.'

Yabba jumped to his feet and whacked the tabletop an open-hander. 'What the bloody hell are yer bitching about, Les?' he bellowed. 'If you can't talk sense, shut yer bloody trap. I called this swarm to find out why the meat is off – that's if it *is* off. We're not here to spear Jack.'

Terry defined firmly, 'I ate two chops, and the third one was just on the turn – so I left it. It might not be that the cook is at fault: this hot, muggy weather turns meat quickly. The presser said he saw a station hand hang a sheep in the meat-house about sundown on Friday. I reckon Jack cut him up at dawn bust on Saturday morning and put the mutton in the fridge. If the fridge is working okay the meat should be right as rain on Monday.'

'That's right!' Scrubber snapped. Belittled by Yabba's chastisement, he bounced to his feet, keen to regain self-esteem. 'If the f—in' fridge was okay the meat would be as sweet as virgin's piss,' he stormed. 'It ain't. The chops are f—in' rotten. I move we sack the cook and get an extra fridge.' Looking around for support, he added, 'Anyway, where I come from the contractor always carries a beer fridge for his men.'

Yabba's day had been hard, and his wick was burning short. A stickler for rules and respect, he said authoritatively, 'I'll tell you this: I measured the fridge before we signed on; there's not much room to spare, but it does comply with Award conditions. But it won't freeze if some of you coves keep putting beer in it. Give the cook a fair go! Keep your beer out of the fridge.' He glared at Scrubber. 'Yeah, I've shorn for contractors who carry a beer fridge. Good on 'em – but it's not in the Award.'

Scrubber stepped up and tried to eyeball Yabba. He whacked the suffering tabletop and declared, 'I say we sit down till we get a beer fridge.'

I laughed. 'Sit down yerself, shorty! Do you have to jump up like a jack-in-a-box every time you want to say something silly?'

Scrubber snarled at me, 'Yer wanna make me sit down?!'

'Nah! You've got duck's disease, mate: your tail is so close to the grass nobody knows if you're sitting or standing, anyway.'

Everyone laughed except Scrubber, who glared at me and swore. The tension eased, and Yabba took advantage. 'Righto! I'll see Brian and tell him we want an extra fridge early in the morning. Now, let's go and get some shut-eye.'

'Hold it!' Frankie snapped. 'It's a waste of time asking for anything without some industrial muscle. Scrubber's got some guts – and I'm with him. I second his move that we go on strike till we get another fridge.'

Yabba roared, 'That motion is out of order and you know it! You've been in the game long enough to know the Union will only support the Award; and if yer want to get down to tin tacks the Award says we can't even bring grog on to a property.'

'The Award is up to shit – the same as the Union,' Frankie said vehemently. 'The Award should have gone out with the blades. It also says we can't bring bull camels or stallions on to a property; and it also says black fellas and Chinamen can't hold tickets, but I've shore with both chinks and coons, and the organisers have sold 'em tickets. AWU stands for Australia's Weakest Union; it's as weak as piss and it's only there to sell tickets to give cushy jobs to bludging officials.'

The Union rep ordered, 'Wake up to yer bloody self, Frankie. I was in this game before the war. If it wasn't for the Union you'd be slaving for a handful of silver and you'd still be sleeping on straw. Instead of bitching about a beer fridge you'd be eating salt meat out of a barrel and spreading mutton fat on a hunk of damper.'

All around argument broke out. Yabba shouted that the meeting was closed. He took his committee men and we went to discuss matters with Brian.

As the team made for the shed on Tuesday morning two station hands unloaded a second fridge and carried it in to the kitchen. They filled the tank with kerosene and trimmed and lit the wick. A sheep had been killed the previous evening, the meat had been hung overnight and Jack had cut it up before sun-up. Before midday the new fridge was beginning to freeze, and Jack packed it with fresh mutton from the overloaded first fridge.

During the afternoon smoko break the rep and his committee went with the overseer to check that the fridges were operating to the cook's satisfaction. Bottles of beer occupied the coldest section beside the freezing chamber in the new fridge. 'Who put this beer in here, Jack?' Brian demanded.

'They left them on the table; I put them in the fridge,' Jack said diffidently. He didn't mention names as he went on, 'Last week the fridge wouldn't freeze with beer in it. It might be alright now we've got two fridges.' It seemed he didn't want to disappoint anyone, and avoided confrontation.

'It won't be alright,' Brian snapped, quickly removing six bottles and standing them on the floor. 'This fridge is hardly cool; it's overloaded – no wonder the meat won't keep. There's to be no beer in the fridge, Jack. Understand! Make it clear to anyone who brings beer in to the kitchen.' He turned to Yabba. 'You tell the men, Yabba!'

'Too bloody right, I will!' Yabba wasn't the man to avoid confrontation.

It was a hot afternoon – about 115 degrees Fahrenheit (forty-six degrees Celsius) in the kitchen. Jack had draped both fridges in hessian which he wet every so often from a bucket of water. It was all extra work. He sat down heavily and put his head in his hands; he looked old and grey and sweat trickled down his face.

'Take it easy, Jack,' Terry said softly. 'Just tell those toe-rags what the boss said. It's nothing to do with you.'

Yabba called a stop-work swarm on the board. 'The overseer just took six bottles of beer out of the fridge,' he reported. 'The fridge wasn't freezing. I dunno why the babbler allowed anyone to put beer in it. Last night we agreed to give the cook a fair go! Loading his meat fridge up with grog is not a fair crack o' the whip. If the fridges aren't freezing you'll be bitching about the meat going off – so wake up to yer bloody selves!'

Scrubber jumped out of the locks-butt he'd been reclining in, and yelled, 'How about you wake up to yerself, Yabba. They treat us like shit! There are two fridges now. Why can't we use one for beer?'

'You're way out of order, Scrubber,' I said crossly. 'Anyway, why don't you cool yer beer in Terry's ice chest, like most of us?' Terry's ice chest was a frame covered in hessian, cooled by water dripping in a shady, breezy spot.

Turning his back, Scrubber walked towards the wool room and shouted, 'You'll keep, yer smart alec.'

Yabba declared the meeting closed, and Brian rang the bell.

Frankie, Les and Scrubber knocked off twenty minutes early that afternoon, quickly unloaded their bog-eyes and scurried for the exit. Brian pulled Frankie up and warned that knocking off early without consent of the employer was a breach of the Award.

Donning a half-smart grin, Frankie explained, 'Les and Scrubber are suffering heat stroke; and I'm travelling with my mates to see they're alright. They need to cool off in air conditioning.'

They returned in time for tea with half a dozen cold bottles of beer.

The following arvo they did the same. The main course for tea was cold roast leg of mutton with lettuce, tomato, tinned peas and hot mashed spuds. Frankie had eaten a bowl of soup and started on the main fare before he asked pointedly, 'When was this mutton killed?'

Working in the kitchen, Jack didn't reply, so I said deliberately, 'It's Monday's kill. Jack threw out what was left of Friday's mutton. I saw one of the station hands hang another sheep in the meat-house tonight.'

'He can throw this one out, too,' Les said, shoving his plate away. 'It's bloody rotten. The mess will cost a fortune; the cook's chucking out more meat than we're eating.'

'Les is right,' Frankie backed up. 'We'll have to sack this bait-layer and get a cook who can put on a decent feed. I can't shear on soup and bloody salad.'

Jack had heard enough, and again he retreated from the kitchen. I resumed eating, while Scrubber, his appetite primed by a day's work and his wits dimmed by two bottles of beer and a couple of shots of over-proof rum, hadn't stopped. Suddenly getting the gist of things, he shoved his plate away. 'The meat's crook again,' he declared.

Yabba glared around the table. 'Jack has put on a bloody good feed. What do youse blokes want – another bloody swarm?'

'Nah! What's the use?' Frankie grumbled. 'You fellas cop shit and like it.'

'That'll be enough of that kind of talk,' Yabba warned, barely keeping a lid on his anger. 'I'll get Jack and we'll check the meat in the fridge tonight. You better have a gander, too, Frank.'

Frankie laughed sardonically. 'No thanks, Rep. You can eat yer stinkin' chops for breakfast. I'll shear on a plate o' cornflakes. We'll go to town and get a feed at the Greek's tonight.' He left with Les and Scrubber in tow.

Under the Award cooks were not required to prepare morning smoko beyond an urn of tea, and men took their own smoko to the shed, however Jack, in the manner of many babblers, made up a nourishing smoko tray of sandwiches, brownie, left-over chops and half oranges.

The following morning, as I trotted to the kitchen at twenty past nine to pick up the smoko, I saw the cook standing in the friendly shade of a tree, apparently inspecting the Austin Champ and its cargo of camping gear. The previous afternoon, after knocking off work a little after five o'clock, I had noticed him in the same spot. 'Hi-ho, Jack,' I had called, but Jack was too preoccupied to respond. Now I called, 'Hey, Jack! Is the tea made yet?'

Jack looked startled. He followed me into the kitchen and made the tea. 'Do you need a hand to carry the smoko?' he asked. It was his usual offer, and I made my usual reply: 'Nope, I can handle her.' And I set off, cheerfully singing 'Pretty Redwing', two smoko trays under my left arm and the urn swinging from my right hand. Zulu stayed behind, trusting the cook would coddle him with a few tidbits. Besides, it was a lot cooler relaxing under the huts than in the shed.

Around eleven o'clock I was glistening sweat as I swung the lever, when the dog raced into the wool room, agitated, whimpering, demanding attention. 'What's a matter, boy? What's a matter?' I asked, trying to console him. Settled by my touch and voice, Zulu found refuge in a recess among the stacked wool bales while I worked on. At a quarter to twelve I washed up, pulled a shirt on, picked up the smoko gear and strode for the mess. Nearing the kitchen I noticed the dog had stopped halfway. I called and whistled, but Zulu lagged warily. I carried the urn and tray into the kitchen.

As ever, Jack's kitchen looked spick and span, ready for the lunch-time rush. Two legs of roast mutton and a tray of baked veggies retained warmth in the half-opened oven, the big iron kettle was steaming, and a saucepan had hot water simmering in it, ready for the unopened cans of peas on the table. I called and cooeed for Jack, but got no answer.

The square meat-house was a few yards from the kitchen's rear entrance. The upper sections of its walls were gauzed for fly-proofing and ventilation. Unexpectedly, the door was wide open. The white painted interior widened in view as I approached, but my steps faltered, and I tried to disbelieve the shocking horror my vision revealed. Jack was sideways to me, sitting on a kitchen chair, his shattered head lolling on his left shoulder. Hair and blood and pieces of his skull spattered the ceiling and gauze, and blood had soaked his shirt and trousers and dripped on to the floor. Half a bottle of beer was before him on the cutting bench, and the shotgun from the Austin Champ rested between his legs.

I turned and ran.

The shearers were emerging from the woolshed, pacing to make time for a feed and a rest before they bent their backs and pulled into gear again. I ran past the dog and halted the rep. 'Hold it!' I gasped. 'The cook shot himself! He's in the meat-house!'

'Sh-shot himself! Sh-shot himself?' Yabba stammered. 'Waddaya mean?' But I had run on to report to the overseer.

Slow to comprehend my terrible news, Brian placed the tally book on a wool table, and queried quietly, 'Shot himself? Is he dead?'

'He's dead, alright,' I replied shakily.

'My God! Tell Yabba not to let anyone disturb anything,' Brian directed, and strode swiftly towards the nearby homestead to report to the owner and phone the authorities. I hurried reluctantly back to the scene of horror. Recalling the upset dog's rush into the wool room, I pictured Jack fondling a farewell with Zulu before taking his final action: 'You're my good mate, *the only one who understands.*'

I had spoken with Jack more than anyone else, yet I hadn't understood. I hadn't picked up at all on the man's struggle with despair as he was overwhelmed by the pain of living, and the peace of eternal sleep beckoned. I couldn't associate the lifeless, shattered corpse with the living, functioning being I had spoken with in the morning: a tired middle-aged man, yet retaining energy, memory, responsibility, care.

I stood back, while Yabba, suffering none of my paralysing melancholy, took control. 'No one goes into the meat-house until the police get here,' he commanded. I wondered if the old digger had been desensitised to violent death in his youth as comrades and Japanese alike were butchered by machine-gun fire and shrapnel blasts, and cooked alive by flame throwers.

Impelled by curiosity, the men looked into the death-room before they assembled in the kitchen and mess. Four hours of hard labour had run down their energy, but only a few chose to eat, particularly after someone surmised that the cook, in his madness, might have poisoned the meal for revenge.

Yabba's response was to load a plate with roast mutton and veggies. 'As soon as the police and the doctor leave we'll have a swarm,' he informed between bites. 'I told Brian that out of respect for the deceased there will be no more work today.'

The police and doctor and ambulance arrived shortly, and Brian and Yabba gathered the dead man's belongings under the watchful eye of the sergeant before they joined me in making statements. Without haste or delay the body was stretchered on to the ambulance and driven away.

The overseer directed the rousies to clean up the dreadful remains in the meat-house. They refused. Yabba took up their case. 'No bloody way, Brian! The boys won't take it on, so it's up to the station.'

Brian addressed the meeting. He had arranged with the police for Jack's earnings and possessions to be forwarded to his next of kin. The funeral would be at ten-thirty the next morning, and work would begin after dinner at 1pm.

'We won't work without a cook!' Frankie objected. 'What's the score there?'

'There'll be no hold-up on that score,' Brian explained. 'We've arranged for the station to do the cooking. They'll bring meat.'

'We can work with that,' Yabba said. 'But we don't want the meat-house used at all.'

'I'll pass that on,' Brian said. 'It should be no trouble.'

The meeting closed, and the team headed for Cunnamulla, while station hands attended the grim reality of cleaning the remains of Jack's shattered head from the gauze, wall and concrete floor. However, despite disinfectant, bucket, broom and hose the imprint of his personality remained to haunt the psyche of the team – and perhaps the vicinity – for years afterward, as the story was told and retold. The meat-house was never used again.

The station cook arrived early the next morning to provide cereals and grilled chops and eggs at eight o'clock. After breakfast, the men shaved, shone their shoes and extracted ironed shirts and dry-cleaned trousers from the bottom of their ports.

Yabba had me drive him to the police station before they joined the team at the cemetery. The police had discovered from evidence found in Jack's wallet and a phone call to the Department of Repatriation in Brisbane that the deceased was a war veteran, classified as eighty per cent totally and permanently incapacitated.

Beneath a pitiless climbing sun, men stood, heads bowed and hats in hand, as Yabba emotionally and emphatically farewelled a fellow World War II veteran. 'Jack was eighty per cent TPI! Totally and permanently incapacitated! The doctors say Jack shouldn't have been working. But Jack wanted to work! And a man's got a right to work if he can do the job. *Jack could do the job* – and he didn't want to bludge. He was in the front line when his country called him. His country – and his workmates – should have taken better care of him.'

I had felt numb and unresponsive since discovering Jack's body. I listened while God's representative said a few worn words about honouring unsung heroes, and their guaranteed reward in the hereafter. The coffin was lowered, and the falling sods drummed Jack's last post.

If Frankie and Les felt burdened by the anxiety they had caused Jack over his last few days, they were determined not to reveal it. A few soft eyes, however, were disguised by coughs and sniffles to hide unmanly displays of emotion; and Scrubber hurried off among the tombstones, sobbing like the child he was. Terry followed him, offering commiseration, but Scrubber wheeled and turned him back. 'Can't a man go for a bloody piss?' he blubbered.

When the team assembled near the cars, their decision was predetermined. 'They must be jokin' if they think I'll be on the board at one o'clock,' Frankie snapped.

Terry measured his words. 'Our comrade died yesterday. We take a day off out of respect. That is today.' By nods and monosyllables the decision was unanimous.

Brian was waiting twenty yards away. Aware that the death of a workmate meant the team traditionally took the day of the funeral off, the overseer had anticipated the decision. Yet, he was anxious: the cost of the shearing was climbing daily, and the station had demanded a one o'clock start, which he doubted he could deliver.

Fronting him with Terry, Yabba declared, 'We're taking the rest of the day off, Brian, in respect for Jack. It's about all we can do for a workmate.'

'You've already had a day off: midday yesterday till one o'clock today. That's a day in my book. I'll ring the bell at one o'clock. Tell them any employee who isn't on the job will be in breach of the Award.'

Ringing the bell was a matter of procedure, and meant the team might be illegally on strike if work didn't begin. Although the Award made no allowance for time off in respect of death, a prosecution would be poor publicity for the company; and a wily defence advocate before a sympathetic magistrate might establish a precedent for custom over law. A magistrate might even grant exemption on the grounds that the team's absence was the result of an 'Act of God'. On the other hand, if the company sacked the team the Union might take it to court for illegal dismissal.

Hard years battling bosses on the shearing boards and confronting army officers had nurtured Yabba's scepticism of authority and his bush-lawyering skills. He was rocky-faced as he declared, 'We are not on strike, Brian. But we are taking compassionate leave.'

Rejoining the mourners, Yabba explained, 'The boss is going to ring the bell at one o'clock. We'll have a couple of drinks to pay our last respects to our workmate, and then we'll all go back to the shed for dinner. We'll have a swarm to endorse our decision, and wait on the board till after the boss rings the bell. This show could end up in court, so we don't want to leave 'em room to claim we were at the pub when the bell rang.'

They quietly raised their glasses in respect for the dead. There were no adverse comments when Yabba proposed, 'Jack was a good cook, a good bloke and a brave soldier,' for no one felt like lifting the lid on their sense of collective guilt.

'Time to go, boys,' Yabba said after a couple of beers. 'We've got to give the cook a fair go.'

*

The long weekend dragged. After lunch on Saturday Scrubber's Beetle revved for town with the usual booze-heads aboard, while others killed time around the huts and shed, spine-bashing, playing cards, reading and catching up on personal chores such as writing letters and washing and mending clothes.

The boozers were home for tea and an early night, but they saddled up for the Sunday morning session and didn't return for lunch. After eating, I took advantage of a summer breeze. Lying in my swag-wrap on the ground in a shady spot, I stretched out reading.

Scrubber turned up around two o'clock, half-shot and on the prod. He stood over me and niggled my ribs with his shoe, snarling, 'Get up, yer know-all bastard!'

'Piss off! Go back to the pub, Scrubber, before I smack your arse,' I enunciated patiently. 'You're not big enough, you're not good enough, and as per usual you're as drunk as a skunk.'

'And yer as weak as piss, Presser. Yer all mouth and no guts. Same as yer were in Goondiwindi. Get up, or I'll kick yer!'

Terry was sitting on his drum reading, a few yards away, and the dog was snoozing beside him. The row woke Zulu, and he sprang to all fours, alert and growling. Terry seized his collar as I rose swiftly. 'If you insist, Scrubber. You go first – out under the pepperina tree.'

'I'll take the f—in' slack out of you, yer big-noting bastard,' Scrubber promised, and began to walk the plank over the bore drain. He was halfway across when I heaved him head-first in to the hot water with a leg throw, and jumped into knee-high water. 'By crikey, Terry, it's bloody hot!' I called, laughing as I ducked the gurgling, struggling Scrubber for the third time.

On his feet, Terry shouted, 'For God's sake, pull him out! The water's too bloody hot! You'll boil him before you drown him.' In the excitement the raging dog broke loose to back up his master, and made a flying leap into the fray. Still laughing, I released Scrubber and grabbed the dog's collar. Back on his feet, the tough little shearer was spluttering, gasping and cursing, but he tackled low and took me down, while the dog swam ashore. On top now, his berserker fury fuelled by vengeance, Scrubber pushed me to the bottom and held me there. Bigger and stronger and over confident, I had taken the combat lightly. Suddenly I realised I was on my back under muddy water, blind and breathless, and unable to find enough purchase in the soggy bottom to dislodge my assailant. I was close to panic when Terry splashed in and hauled Scrubber off.

'Cripes! It *is* bloody hot,' the big shearer cried. As he turned for the bank, Scrubber nailed him with a right swing before going after me. I had scrambled to my feet. Coughing muddy water and half-broiled, I measured Scrubber through blurred vision and punched viciously with both fists. Scrubber dropped, and I ducked him vengefully three or four times, while Terry wiped blood from his eye and shouted, 'Drown the thankless little bastard!'

We laughed in the shade of the pepperina tree and chiacked Scrubber as he scrambled out, belching water and nearly spent. He sat on the bank for a couple of minutes.

'That should've sobered you up,' I called. 'Are you alright, you poor little chap?'

'None the better for your asking, yer stupid c—,' Scrubber spat out, and walked to his VW.

Terry called, 'I've got a mind to throw you back in the bore drain and wash your mouth out.'

'You and what army? It took two of youse big bastards to beat one little bloke.' He turned and put up his fists. ''Ave a go now, yer smart alecs! One at a time, I'll take youse all on.'

'Yer a glutton for punishment,' I said, and laughed derisively as I strode purposefully towards Scrubber. The little shearer scrambled to safety behind the wheel of his V-Dub, reversed, revved up and charged. Terry and I jumped for cover behind the pepperina tree. Scrubber did a couple of victory wheelies and headed for Cunnamulla, shouting abuse and insults out the window, emphasised by the upturned finger.

The Kahmoo shearing cut-out suddenly. Terry, Carl and Yabba shore fair tallies on Monday morning, while the learners and the grog-afflicted struggled in the low twenties. Scarcely exchanging a word, the team dolefully ate smoko in the wool room. Even Yabba didn't care to raise a conversation. We had resumed shearing for only a few minutes when the bell pulled us up. I was lying on my back, reading while I waited for the wool to build up. I guessed a mechanical breakdown was the problem until Brian stood above me, grinning like a Golden Casket winner. 'That's the last bale, Presser,' he declared. 'The shed is finished. KAPUT! No press-up. Leave the wool in the bins – it's all over, drover!'

I was surprised it had lasted as long as it had. There had been more strife than you could poke a stick at, and tallies were a long way below par. 'Fair dinkum, Brian?' I asked, rising swiftly. 'Are we sacked?'

'No, the team is definitely not sacked,' the overseer explained pleasantly. 'I've been instructed to close the shed. If you report to the Charleville office it's on the cards you'll be placed next week.'

I realised the term 'sacked' wasn't used because the AWU might sue the station for compensation for wrongful dismissal; and a sympathetic magistrate, on hearing the team had followed accepted precedent in taking a day off in respect of an ex-digger's death, might find in the Union's favour. But UNGRA could split hairs and find an excuse for 'closing the shed' – if summonsed. Meanwhile, I would be glad to collar my cheque and roll my swag.

15

THE LAUGHING KIWI

The Charleville 'first half' run finished prior to 30 June to convenience grazier's tax returns. I enjoyed a brief break with family in Brisbane before driving north with my mate, the Laughing Kiwi, to join Richie Sack 'em Jack's team for the 'second half' run. Richie's final shed cut-out about fifty miles from Hughenden at five o'clock on an afternoon in December. Doug gave me a hand to press-up. Always as impatient as a hungry cattle dog penning up meat pies, he yelled, 'C'mon! Shake a leg-iron, you convict bastard! They'll drink all the piss before we get to town. We can shower at the pub. Let's collar the cheques and hit the road.'

After a few ales and a cafe feed, we phoned 'Bullshit Bill' to line up the next year's employment, and said farewell to a few mates, lively barmaids and a couple of publicans who could be relied on for accommodation and a loan if we were ever short of a quid.

We were aiming for a 5am take-off. Doug said, 'You'll be on deck at sparrow's fart so dig me out at a quarter to five and we'll just get in the car and go. I'll fuel up and grab us a couple of pies at Longreach while you phone your mum. We'll be having a beer with Jack by eight o'clock.' He laughed, and added, 'And tell Pat her favourite boarder will have snags and mashed spuds and peas; and don't forget loads of gravy.'

I grinned. Dad was at home, having finished a shearing run around Charleville. He and Doug were natural mates, while Mum had a sparky relationship with the young shearer. His dark curly locks, baby-faced smile, infectious warmth, respect for family convention and handiness with a tea towel seemed to bring out the maternal instinct in most females, but Mum was always wary of him, and her Catholic conscience compelled her to pray for his moral reconstruction. On 'blue' days she suspected he was an over-sexed Lothario, leading her son astray and urging him along the highway to hell and damnation in an eternal afterlife.

Driving to West End in fifteen hours was a game estimate, typical of Doug's optimistic attitude. The 900-mile trek would include some 600 miles of dirt, or single-lane dagger-edged bitumen decorated with potholes. Canny travellers carried a tucker box and plenty of water because a seasonal storm could pickle your plans and maroon you for days – forcing you to camp rough roadside or in a pub if you were lucky. We had made it before, though, with the hammer down and a carton of longnecks, plus a bag of meat pies.

We'd need the luck of the Irish as Queensland roads harvested 500 souls a year. Boozy lead-foots would drive over dirt tracks and through wash-outs and dust-clouds, out of which ambushing roos, bullocks and cattle trucks charged blindly, dealing chaos and death and destruction. The media dubbed the bush roads 'crystal highways' because of the fragments of a billion broken bottles and busted windscreens littering the roads and table drains.

Heedless of the fearful statistics and fuelled by the optimistic adrenaline of youth, we drove flat-strap through Queensland's killing fields under the auspices of the three Saint Christopher medals my mother had installed – one in the glove-box, one swinging on a leather boot lace off the choke, and another a permanent resident in my wallet.

Walking back to the Western Hotel, our accommodation for the night, I was lost in thought of a joyous family reunion. I didn't notice the Laughing Kiwi's unusually sombre silence.

Entering the pub, I stepped towards the stairs, but when my mate aimed for the bar I followed him and ordered a tomato juice. 'What's up, cobber?' I demanded anxiously, as Doug gulped a double rum and started on a beer chaser.

It was a bad omen. The Laughing Kiwi was habitually a happy beer drinker. On the surface he was a hard-working hedonist who shore sheep only to fund the good times, but on rare occasions he'd have a bout of the blues and his doppelganger would emerge and demand rum.

I'd first run across Doug at a shed I was pressing at in early 1958. He was there with his brother who looked just like him, but while Doug was full of fun Bob was burlier and surlier – at times provocative.

The two young shearers, from Huntly in the North Island of New Zealand, had ventured to south-west Queensland early in 1957. Unlike most Kiwi shearers of the time, they had persevered with the narrow-gauge combs and cutters and the challenge that tougher Australian merinos presented. With plenty of aptitude and application the boys were 'getting their average' within twelve months, and some of the older hands reckoned they had enough sheep-shit on the brain to become guns in a couple of years. With Award rates in Australia far superior to New Zealand's, as their tallies and earnings went up the boys thought they'd struck gold.

Cashed up, they bought a new VW, and between sheds and on long weekends they drove 300 miles to rev and rage up and down the Gold Coast. Sand and surf held no attraction; they boozed and partied, and the hunt for exciting, available femmes was the great game.

On their final excursion Bob and another man's wife displayed obvious mutual attraction. Instead of opting out when the husband became jealous and intervened, Bob became provocative. In the ensuing all-in brawl the burly shearer was stabbed in the heart by a small knife blade. The jury agreed the husband acted in self-defence – and he walked.

To outward appearances it seemed that Doug had donned a brave face and moved on. The few who knew him better, however, understood that in his mind justice had not been served and his inability to square things haunted his subconscious. Broken-hearted, depressed and drunk on rum, Doug would begin whimpering, 'If only I'd been there Bob would be alive. I should have been with him. He didn't know when to back off, but I could calm him. I could talk sense into him. I promised Mum I'd look after Bob. He was older, but she knew I was the one with common sense. Bob was Dad's favourite, but he don't ever mention him. It broke his heart. And I broke their hearts. I let them down . . .'

He would curse the killer and swear vengeance, and then sob at his impotence. And on two occasions, after his anger burned out, I held him in my arms in our guest room at the Grand Hotel as he cried like a confused and helpless ten-year-old. Feeling inadequate and embarrassed, I wished Doug's mother would materialise; or that any motherly woman would walk through the door and succour a child in misery.

In the pub, my mate was staring sullenly into his third double over-proof rum. I knew from experience that Doug's doppelganger would not withdraw until he had harrowed his host with hours of remorse and self-degradation; that the Laughing Kiwi would turn to threatening revenge against his brother's killer before the finger of accusation turned inward, causing him to writhe under the terrible guilt of failing his beloved brother who was almost a god in his young life. I was expecting to give a long counselling session, talking quiet words of support and encouragement, when Harry the Spiv boisterously belted me on the back. 'Just the pair of suckers I wanted to see!' he said. 'You'll make up the number for a game of poker in room eight.'

'I won't,' I said firmly, 'but Doug will.'

Doug swallowed his rum and turned around slowly. 'Too correct he will! Any old time at all I'll laugh and take your money, Harry.'

I rolled my eyes upwards to silently thank Fortuna, my favourite Goddess, and said to Doug, 'Lay off the rum and I'll stake you twenty quid. My loss, and halves if we win.' I considered it an investment: Doug rarely lost, and the game would likely get him off the rum and banish the self-loathing that was feeding the blues.

'You're on,' Doug said. He collected half a dozen cans of VB, pocketed the two tenners and straightened his shoulders.

The Spiv fancied himself as a woman-winner. He leant over the bar and confidentially engaged the barmaid. 'Gloria, darlin',' he said, 'here's a quid. Be a good girl and fetch a plate of cheese and crackers to room eight when you close up. And give me a kiss in advance and you can keep the change.' She blew him a raspberry as she snatched the note. 'Make it a fiver, Spiv, and you can kiss my ring.'

Confident my mate and my investment were safe, I hit the sack. Hours later I heard Doug crowing, 'We won sixty smackers, mate.' He fell into bed, chuckling, but he was feeling seedy when I tried to haul him from the sheets at seven o'clock, and he fired a battery of bad language. I had packed the car, run the dog, eaten a hearty breakfast and read for an hour, when Doug arrived at nine o'clock – his grin restored by a shower and a double brandy reviver. 'We are a wee bit late,' he observed. 'But we can still make it to Brisbane tonight.'

'We would be hours along the track if you'd pulled your arse into gear.'

'Tut-tut, me lad! You're only crabby on account of I collared sixty smackers for us while you were snoring.'

I said curtly, 'There's not much change in nine hundred

and fifty miles from here to the big smoke, so let's hit the dry and dusty. Starting this late, Roma will do us tonight. The old folks at home will have the midnight horrors if they think we're flogging down the range in the small hours. I'll phone Mum when we hit Mitchell.'

Zulu usually lorded it over the back seat, but Doug was having none of it. 'Over the front, you biting blue bastard,' he ordered. He doubled his legs and stretched as far as the confines of a VW back seat allowed and nodded off.

At Longreach, we downed a few beers and ate meat pies for lunch. Despite Zulu's grovelling and dribbling he was denied a pie because, as Doug put it, 'Give that dog a pie and he'll fart fumes that would stop my grandmother talking – and the old lady can gab under water with a mouthful of cornflour.'

We pulled in at Tambo for a couple of 'quick ones for the road', but were cheerfully delayed by the arrival of two girls who were returning to Brisbane for Christmas holidays after a year teaching in a school at Mount Isa. They had been on the road for eleven hours and had pre-booked a room to overnight in Tambo. Confident the teachers would benefit from our company we bought drinks and offered dinner, and the four of us took seats in the dining room and chatted casually over the menu.

'Looks like a feast fit for a king,' Doug said. 'Roast beef or Monterey cutlets, baked veggies and greens.'

He grinned. 'Beef and plenty of greens for me! Whether it's Monterey or Timbuctoo or Sydney, it's all mutton to me. If I eat another chop I'll start growing wool.'

'That would be an awful shame,' said Janice, the shorter, brunette teacher, in mock sympathy. Running her fingers through his black curly locks and smiling archly, she added,

'With those chubby pink cheeks and lovely curls I'll bet you were your mother's favourite, and the teacher's pet.'

It was an opening made to order for the Laughing Kiwi. Engaging Janice with his dimpled, cherubic grin while taking her hand, he sang in a soft melodious baritone:

> I must have been a beautiful, baby;
> I must have been a wonderful child.
> When I was only startin'
> To go to kindergarten
> I must have sent the little girls wild.

I grinned. Doug's 'lead-in' might seem to be as preposterous as it was spontaneous, but I'd seen my baby-faced mate work his combo of charm and song several times – and on occasion it enticed a 'starter' to become a 'sure thing'.

'We shall have to wait a while,' Marion, the tall fair English teacher, expostulated. 'The dining room is filling up and there's only one waitress for all these people.'

It seemed her words summoned a second waitress. Unnoticed behind Marion and me, she attended patiently as I bandicooted Henry Lawson for an attention-seeking declamation:

> What's that, waiter? Lamb or mutton! Thank you –
>   mine is beef and greens.
> Bread and butter while I'm waiting. Milk? Oh, yes – a
>   bucketful.
> I'm just in from west the Darling, 'picking-up' and
>   'rolling wool'.
> . . .
> All day long with living mutton – bits and belly-wool
>   and fleece;

Blinded by the yolk of wool, and shirt and trousers stiff with grease.

. . .

Picking-up for seven devils, seven demons out of Hell!
Sell their souls to get a bell-sheep – half-a-dozen Christs they'd sell!
I said beef! You blood-stained villain!

. . .

Beef – moo-cow – Roast Bullock – BEEF!

Doug and Janice laughed aloud while Marion held a tight rein on an amused smile. Waiting for quiet, she assessed primly, 'At best a crude piece, reflecting the standard of much of the colonial taste in literature of the period. But timely and well presented, I must admit.'

An intellectual stuck up snob, trying to put me down, I thought, slipping into defensive silence until a laughing Janice caught on: 'School's out, Ma'am,' she called. 'Tell that boy he can come out of the corner and write one hundred lines.'

I joined in the laughter, and Marion jabbed me in the short ribs. 'School's out,' she instructed. 'You must go and write one hundred times: *I must not take everything seriously.*'

The waitress waited for the laughter to subside. 'Another waitress will be here soon,' she said. 'I have telephoned her that the rush is on. I am the cook, but I will help out until she comes.'

Doug said, 'It reads here, "Chef's Special". If you are the babbling brook – who is the chef?'

A puckish smile played around her lips and dark eyes twinkled. 'I am the babbling brook, and also the chef, but

I do not have the paper qualification. My mother taught me to cook. It is Hungarian home cooking. No food is tastier or more nutritious. Your orders I will take now, for other folk are waiting.'

She squeezed my shoulder and bicep and said appreciatively, 'This poet is not like other poets I have known. They are usually pale and skinny from struggling with the conflicts of conscience and desire. But his young brain and body will benefit by the Halaszle, my own fish soup. Here, I create it from the fresh yellerbelly. I must admit it is the equal of what I made in my home land, but not as spicy. You Aussies do not appreciate the *hot stuff*.' Her puckish grin dimpled her face again as she continued, 'The goulash is beef and paprika, simmered until "the cows come home" – like you Aussies say.'

We laughed with her before she added, 'It is served with dumplings and, if you like, baked spuds and greens.'

'You've won me, honey!' Doug exclaimed. The teachers were delighted at the unexpected discovery of continental cuisine in Tambo. The cook took an order for specials all round, and returned to kitchen duty.

'Bloody good ky!' Doug said, polishing his plate with a hunk of bread.

'What is ky?' Marion queried.

'A Maori word for tucker,' Doug replied.

'An articulate example of vulgar terminology,' Marion declared. 'But I find myself in absolute agreement: this is *bloody* good ky.'

The girls ate sweets and finished their bottle of wine while the boys had another beer; then they caught each other's eye, and thanked their hosts for 'a pleasant and entertaining evening'.

'We are the early birds,' Marion said. 'It's after seven-thirty, and we are up and away at four am.' Janice spun around at the doorway and hit Doug with a wave and a wink. He was on her in a flash, and they engaged in a brief but enthusiastic kissing cuddle, before she pushed him away. 'You *are* a naughty boy,' she said, and fled, giggling, with Marion.

Back at the table Doug said, 'Gees! *School teachers!* They shouldn't even be allowed inside a classroom! They don't know the basic rule: a bottle of red and a good feed is worth a root anywhere.'

I fed Zulu and returned to the bar. The place was buzzing with drinkers: shearers, ringers, truckies, council workers. A loud, beefy foreman put a five-pound note on the bar and challenged all-comers to an arm wrestle. There were no takers as the locals knew the power of his great bicep too well, until Doug stirred the possum by throwing a ten-pound note down. 'You're on, bro, best of three, ten quid,' he challenged.

Doug was a work-fit, nuggety twelve stoner. The foreman, who was carrying a fair bit of flab, stood up to reveal he was six feet and probably scaled eighteen stone. Looking down, he grinned derisively as they shook hands, clearly thinking this little man would be just 'meat and drink'. I knew he was in for a shock. I had seen my mate win the championship in his home town of Huntly, New Zealand, by besting a Maori strongman; and I'd seen him outstay all takers from Cunnamulla to the Cloncurry.

The Kiwi took his time warming up his muscles; then grabbed the attention of the crowd and a round of applause and a few boos with a one-man version of the Haka.

I worked the punters for side bets. The drum I'd heard circulating was that the local man's only moral support came from a few of the foreman's arse-lickers – sycophants who basked in his fat shadow. The punters would like to see him cut down to size, but they didn't think that the Kiwi had the heft to do it. I had deposited a hundred quid with the barman to cover my losses – insurance the punters would be paid if Doug lost – before I pocketed their bets and wrote the details in my notebook.

I had laid sixty quid at good odds when a tall, weather-beaten sheep cocky shook my hand. At that time many graziers were still living high on the hog of the 1950s wool boom – you could pick 'em because they had handshakes like bank tellers – but hard muscles and a strong grip verified that this bloke was still on familiar terms with the 'Spaniard', also known as 'Manuel Labour'.

'Look, young fella,' he said urgently, 'I'd bet you fifty quid, but I don't want to take your money. You fellas are shearers, aren't you? I served my time on the board in the fifty-six strike, and I reckon you blokes get it hard enough without giving it away. I'm no Don Athaldo, but I'm no pushover, and "Big Boof" rolled me as easy as a school-boy. He beat Cavanagh for a hundred quid. Do you know Cavanagh? He presses my wool every year.'

Know Cavanagh? Cripes! The giant was a legend; a six foot five inch outback Hercules. I thanked the cocky and immediately closed my 'book'. I'd already bet half Doug's poker winnings, and blokes were still lining up to lay odds on what they reckoned was a sure thing.

The contest took place on a small, round, solid wooden table. I manoeuvred to a position where I could see fair play: the free hand above the table all the time, elbows on opposing sides of a chalk line and a thirty-second break between engagements. The men sat opposite each other, their right hands clasped. The referee saw their grip was exactly above the white chalk line before he ordered, 'Take the strain . . . Go!'

The bar became as quiet as a cathedral as the struggle ensued. In seconds Doug's knuckles were forced to within six inches of the tabletop – but he held, and held, and slowly returned to seventy degrees. He could not regain the perpendicular and the big man regrouped and took him down in just over five minutes.

Doug stood up and laughed. He rolled his shoulders and dropped his arms. Big Boof grinned triumphantly and then swallowed a pot of beer, lapping up praise from his circle of cronies. His reddened face and heavy breathing, however, stoked my confidence.

The ref called, 'Five seconds!' and they clasped. This time they locked at the upright for fully a minute before Doug's forearm swooped backwards. A crony hollered, 'It's all over, drover!' But Doug held at forty-five degrees, before he fought back, ever so slowly, into the champion's territory. The clink of coins and glasses ceased, and only the odd shout of encouragement for the Kiwi came. Doug sat hunched forward, concentrating quietly. Boof was upright, his face flushed, his breath rasping. He looked desperate, but with his reputation at stake he came back with a grimacing surge that carried his nemesis to within ten inches of defeat. Doug held, and I tried to encourage him telepathically with a Kipling mantra I'd often facetiously quoted to stir him on the shearing board:

WOOL AWAY, BOY!

If you can force your heart and nerve and sinew
To serve your turn long after they are gone;
And so hold on, when there is nothing in you,
Except the will which says to them 'hold on'.

Inch by inch the Laughing Kiwi fought back, and buzzing murmur circulated as watchers, despite having money at stake, began to murmur encouragement for the underdog.

Above the chalk line the locked fists halted for a moment – and then Boof's forearm crashed in defeat, his knuckles banging on the tabletop. Doug rose, laughing, flexing his shoulders, while I admonished the ref: 'Thirty seconds, Ref – thirty seconds!'

Big Boof looked buggered. His second snarled into my face, 'You go and get well f—ed. The champ's cramped.' Turning to the ref, he demanded, 'Five minutes time out, Billy.'

Picking him for a big mouth – a gutless wonder without Boof's back-up – I long-armed him out of range with a knuckled left shove in the brisket, and again called on the ref: 'Thirty seconds, Ref – only ten left.'

Glaring, the harried ref snapped, 'I'll see a fair go – so shut up and back off, son!'

He ordered, 'Take the strain'. Boof jumped the gun. It didn't matter: the shearer was primed – and in half a minute took the big man down. The clink and cacophony of the bar-room resumed, and the sheep cocky emerged and quietly slipped me a tenner. 'The wife and I are going home. Give that to your mate. That standover bastard king-hit my brother at the Blackall races.'

THE LAUGHING KIWI

There were plenty of back-slappers and well-wishers volunteering to shout the victor, but the Kiwi only had a couple of beers before he slipped away. 'Early night, mate,' he advised. 'You ought to pack it in, too; we'll hit the track at sparrow.' I wished him luck. I knew Doug would chance a discreet knock on the teacher's door before he threw in the towel.

16

THE GYPSY COOK

I wandered into the lounge where a family night was in full swing. Couples and kids were dancing to the lively music of a wholesome 'Mum'. She was swinging her accordion, while a lean sun-browned singer was hanging ragtime and jazz piano on spirited lines of fiddle music bowed by a fourteen-year-old girl in school uniform. Her dad, a Pommy piano player, called above the music and hubbub, 'She looks eighteen if she gets her own way – tight jeans an' ruby lips an' all. The uniform lets you lads know she's a school kid – and that Dad's in the window!' His laughter failed to mask his pride.

Feeling a bit woozy, I had taken a seat to enjoy the musical scene for five minutes before calling it a night, when a husky contralto murmured, 'I hope you do not mind? I bring a glass of beer for a lonely young man – and red wine for myself. I am Marguerite, and you are . . .

Alan?' I smiled and nodded. Immersed in the betting and battle of the arm wrestle I had forgotten the comely cook. I had been briefly aware of her playfulness in the dining room but, attentive to Marion, had dismissed her as too old. In fact, she soon disclosed, she was thirty-two – nine years my senior.

As she gathered her skirt and sat I was captured by the graceful femininity of a small, shapely woman with a flawless coppery complexion. She had showered and changed into a modest black skirt, creamy top with puffed sleeves and embroidered cuffs beneath a waistcoat of red, white and green. Her long black hair was down, and adorned with two yellow flowers.

When I asked if it was a traditional costume she smiled and tugged at the waistcoat. 'It is the colours of the flag of Hungary – of my native land. I make these clothes to remind me of joy and sadness, and sometimes pride. My shoes are good for dancing.' Sensually, she extended her left leg, drawing attention to a trim ankle, and a neat foot enclosed in a low-heeled black shoe. Well-worn and lovingly polished, the shoe was secured by a silver buckled strap.

I felt sexual desire beginning to ignite.

'I bought them in Sydney,' she said, 'but they are much like the shoes for dancing I wore at home.'

The music changed, and Marguerite laughed and tapped the table in time. 'You keep my place,' she said. 'Soon I will be back.'

Marguerite was dynamic on the dance floor, creating her own world of motion and music, whirling solo among couples with swift-stepping exuberance to a short bracket of polkas.

The pianist announced something I lost in the noise. Marguerite summoned me with an alluring smile and a beckoning forefinger. I was still feeling slightly woozy, but she guided me through a jazz waltz. Then, sitting me down, she took the accordion from the jovial, perspiring 'Mum', nodded to the violin player, and they swung into a medley of fast Gypsy dances while the piano thumped bass time. Two swarthy, solid men in khaki overalls and workman's heavy boots leapt up on cue to buck-dance energetically with intricate practised steps. Applause and calls of 'encore' and 'more' followed what was, apparently, a spontaneous but not infrequent performance.

Rejoining me, Marguerite introduced the khaki-clad men as Bela and Harry, 'my *brothers*, contract fencers . . . from my homeland, Hungary. They were almost neighbours to my father's farm. We were not acquainted. Is so strange: we first greeted the glory of God but fifteen kilometres apart in our beautiful country. We attended the same convent, but did not meet because my *brothers* were old enough to complete school before I began. So strange! We survive wars and hunger and hurt and loss, and then we meet and laugh and dance in – how you say? – "back of beyond".' She laughed – a wonderful infectious expression of thankful joy – and Bela and Harry clasped her in what I hoped were brotherly hugs, before departing to join a two-up game.

Turning back she took my hand. 'Come,' she said softly. 'Here it is too loud to hear.' She led me through the rear door to a table on a verandah.

THE GYPSY COOK

Marguerite listened attentively while I told her about the close bond of my family. I explained how my mother had often said that courage is the first virtue, because only rooted in a bed of courage can the lesser virtues grow; and that while her life had not been easy or financially secure, her courage had been underwritten by the certainty of family love, the surety of self-worth and the guarantee of God's will.

Marguerite nodded. 'Good women are strong,' she said, 'stronger than most men.' It was a statement rather than a supposition.

I had been brought up in a masculine world of competitive work and sport, of hunting and fishing, and found that hard to accept. But I didn't dispute it.

Marguerite continued, 'My mother was strict with my brother, Samuel, and me. Father was always gentle with us, but outside he was a fierce fighter for his beliefs. It was the 1930s, the Depression, and many suffered hardship. Hitler's armies were reshaping Europe, and many people spoke of the benefits Fascism would bring. We had little money, but our farm kept us from poverty.'

She paused for a moment before adding, 'My parents were social democrats. They detested and feared Communist and Fascist ideologies. My brother and I learnt their values. And then our mother died; it was the winter sickness, pneumonia . . .'

A few minutes earlier on the dance floor with bright attire and yellow flowers she had emanated gaiety. Now, she paused and tears misted her eyes. Catching her sorrow, I covered her hand with mine.

'It is alright,' she said, gently squeezing my hand, 'death is a door to the continuation of life.'

The accordion was swinging a waltz. 'Would you like to dance, Marguerite?' I suggested, thinking to cheer her.

'It is kindly, thank you, but I wish to talk, if you will tolerate me.'

'I'm a good listener,' I replied as brightly as I could.

'My country was afraid of Germany's military power,' she said earnestly. 'Hitler would not invade Hungary if we joined the Axis forces. My brother was conscripted, and he died somewhere on the Eastern Front. Millions of fine young men died fighting Hitler's evil war: my countrymen, Russians, Germans . . . They are gone now. Such a waste. Poor boys . . .'

Again she hesitated, swamped by memories, before continuing: 'My maternal grandmother was Roma. Grandfather died and she rejoined her people. I loved my grandmother – Mother and I spent holidays with her people. They taught me to dance and play the accordion. Our government had protected Jews and Roma from the Fascists, and in 1944 we tried to negotiate a separate peace with the Allies. Germany invaded, though, and the Jews and Roma were deported to death camps. My grandmother and her people disappeared. We were fearful my Roma heritage would be found out and so went to Budapest, where we had friends to hide us. Father fought with the resistance; there was much street fighting, and the Germans bombing.'

Marguerite spoke passionately of the hell the people had endured during the bombing and Russian siege, and destruction of her 'beautiful city'. 'The fighting was in the streets and buildings. I hid with women and in the cellars. Always we were cold and hungry. I remember the kids crying. Their mothers hushed them, for we were terrified of being raped and murdered. A priest brought us a box of candles and food. We dressed the wounds of fighters . . . For days at a time I didn't know if my father was alive. No one was safe . . . I am sorry.'

She paused to wipe her eyes, while I tried to comprehend that she was speaking calmly of terrifying events beyond my imagining. Telling of the stricken city's silent welcome to the Red Army, she managed a wry smile. 'It was as if we knew, as you Aussies say, "We were out of the frying pan and into the fire." Many Red Army soldiers raped and thieved. The Communist rulers executed thousands of our people, and many thousands of Hungarians were deported to labour camps.'

I had been enthralled as she spoke of tragedy, of beauty, of good and evil. Now, smiling gently, she took my hand. 'It is good to be with a young man who is intelligent; someone who listens. Even Bela and Harry do not listen. They have become like Aussies – they would rather drink and laugh and talk of work and gambling.'

After relating that her father had decided they must escape to England, Marguerite spoke dramatically: '*Albion*, my father called England; *a place where men can speak freely of truth and justice* . . . The journey was difficult, but my father had contacts in free Europe, and he found a way, and money. Many fled their homeland, for escape would soon become more difficult and dangerous.'

Marguerite told me she had settled to work and study and gain command of the English tongue, qualifying in psychology because she wanted to help people who were 'hurt and shamed by the war'. Leaning forward, she took my hands in her small firm grip, and spoke fervently. 'People are haunted by their memories and dreams of lost loved ones. And they are afraid the madness of gunshots and explosions and turmoil will again turn their world upside down. War reveals goodness and greatness – and evil and depravity. War changes people – we do not recover. Our bodies may look the same, but inside . . . our minds – even our souls – are wounded. You do not understand, Alan? I am sorry. How can I expect you to understand? The experience you must have yourself. For years I never felt safe. People we trusted turned collaborator – or we thought they might. We had to keep moving. I have had to learn to trust again, for to be a worthwhile person, a human being, one must trust. Sometimes I cannot judge – I can only feel.'

Her smile became wistful. 'Eventually I married a handsome carpenter and we emigrated to Australia. My father was already in Sydney, working as a language tutor, and we settled near him.

'I was happy to be with him, and my husband was happy with his mates doing construction work. But we were not happy together. I think it was my fault. My husband was . . . we could not comprehend each other's needs. He did not read, did not think – always with his mates, and laughing and drinking, and work, and work . . . Together we would go to the movies – always his favourites. At first in England I was in love. I was young, but I should have known it could not succeed. I too like to work – to make a commitment – and to read and study and take photos, and eat at restaurants. All of those are necessary for me to be happy.' She laughed. 'Of course, you know I love to dance and party, too!' Becoming serious, she said, 'He believed I loved my father better than I loved my husband. He made me feel guilty, for I knew I did.

'I have been studying and in six months' time I will be authorised to practise psychology in Australia. In time, God willing, I will return to Europe to help. So I decided to see your wild and lovely country while the chance is offering.'

My traditional upbringing had ingrained in me the belief that women were the hub of the home, and their chief duty was to reproduce and nurture. 'Will you have children?' I queried.

She smiled softly. 'I love children. Babies are so wonderful; always they are curious, always clasping with their tiny hands and tasting and touching.' She cradled her arms and swayed, humming and crooning a lullaby. Tears softened her smile, and she said, 'I am sorry. I do not often drink; drink makes me sad . . . I think of all the orphans and the dead children.'

She sobbed into my shoulder and I put my arm around her. I had felt inferior to her knowledge and experience, but she was suddenly vulnerable, and I felt empowered and protective.

Soon she withdrew and dried her eyes. 'My grandmother was a witch,' she said, laughing to lighten the moment. 'She had the second sight. I too have it, but only now and then.'

I laughed. 'I hope she was a white witch.'

'Sometimes she comes to me – like a dream – but it is not a dream. Long ago she told me that white witches are good but they are not all wise, for some will not listen to their dark angels. To be wise a witch must converse with her dark angels, but her angels of light must be stronger or dark angels will dominate her understanding with evil advice. One must take care, for like the Sirens the dark angels are sweet talkers – *like some men*.'

She laughed while caressing my face. 'Perhaps to be wise you must tie yourself to the mast, but you must not plug your ears.'

I laughed with her and she said, 'It is good to joke, but now I do not joke. It is easy if you believe and have faith; but if you have lived in my shoes, and a million other shoes, it is hard to believe in the ruling goodness of the human spirit, for Evil has a human face and Good has a human face. Evil is the eternal liar, a cunning liar, for it often twists the truth into lies. It has great desire to seduce and dominate, and it delights to inflict hurt on the body and despair in the soul. Evil delights in war and chaos. Good believes in truth and in joy, and has moral intent – so it must have faith that can draw on stronger resources than the human spirit . . .'

I went to replenish our glasses, and when I returned Marguerite met me with a merry smile and took my hands. 'You are puzzled; now I puzzle you more. God must have wanted my father and me to live, for he sent strong Guardian Angels. We worked our angels very hard!'

Her throaty chuckle delighted me, and she said, 'When I was little my Roma grandmother took me and other little girls on three occasions to the secret place of her

Forest Goddess. We watched the wood spirits dance and play. We joined in the dances, but we could not touch the spirits, they were so fast and elusive. They looked like little girls. They had flowers in their hair, and they were dressed like us.'

'Was it a dream?' I queried doubtfully.

'Oh, it was no dream!' she replied. 'But my grandmother had warned us: "You cherubs must believe, or you will not see them." I don't think she could see the wood spirits, because she kept asking, "Can you see them?" I wasn't bold enough to ask if she could see them.'

At first I had been amazed that Marguerite was interested in me. I had felt shadowed by her life experience and education, but now her words caressed me like warm waves on a moonlit beach, as she stroked my hand and said, 'Like its beautiful Goddess the moon, love has many phases – more than the moon.'

Rising neatly, she pushed her breasts against me and whispered playfully, 'You are getting tired. Come, it is time for bed.' She said it as though it were the natural order of things and, taking my hand, led me upstairs.

I awoke to the blinking sunlight of wind-whipped curtains and the faint aroma of rose perfume. I hadn't heard Marguerite rise and dress. She must have drawn the curtains closed and departed.

I recalled the heat of my desire the night before as I had watched her open the window and tie back the curtains; then, laughing, push me back and insist we hold hands and watch the stars in silence before embracing. 'There is a garden below this window,' she had said, 'a flower garden. I like to sit there and read. The scent of the flowers takes me to being a small girl, gathering wild flowers for my mother.'

Sitting up, I saw a towel, soap and razor and my neatly folded clothes on a duchess. I smiled to myself and got up and found the bathroom, where I showered and shaved.

I saw no one until I entered the kitchen, and found the voluptuous accordion-swinging 'Mum' sweating over the sink. She looked up and smiled. 'Breakfast's off, dearie,' she said, 'but if you grab a tea towel you can join me in a cuppa in a jiff or three.'

While I worked a tea towel Mum chatted above the clatter of dishes. She was a joyful, hearty romantic, as loyal and defensive of family and friends as a blue heeler.

I gathered the courage to query shyly, 'I thought Marguerite was the cook?'

'Marguerite? Oh, Margie cooks Tuesday through Saturday, and I do Sunday and Monday.'

'Where is she now?'

'Oh, you poor boy! You were with her last night. Don't break your heart; Margie dances every Saturday night and then goes to bed on her own.'

Mum prattled on while she made tea and toast. Marguerite, she said, had puzzled her for six months: she enjoyed Saturday nights, dancing and talking, but remained self-contained and sober and never gave any sign that she might be 'on' with anyone – or even let herself go for a one-night 'play'. She added, 'And with five kids and a horny husband, I reckon I ought to know.'

I tried to hide my embarrassment while Mum opined that she couldn't imagine how a woman could be happy for months on end without a man in her life to love and plan and fight with.

'I'd like to say goodbye to her,' I said.

'She went off with her brothers at seven o'clock. She said she was going to mass in Blackall, and then on to Gem Fields to see friends. Fact is you never know what any of these New Aussies are up to – specially our Margie – she's a quiet one. Still waters run deep.'

The mention of the 'brothers' prompted a fit of jealousy to whack me like a Bob Fitzsimmons solar plexus punch. Mum banged on, punctuating her chatter with belly laughs that jiggled her copious bosoms. 'Great thing about being a Catholic: you can fornicate and gamble and grog on, and Jesus forgives – as long as you drop ten bob in the plate and say you're sorry.' Noticing I wasn't joining her chatter, she queried, 'Are you a Catholic, lad? Me and my big mouth! Anyhow, us proddyhoppers – we just let our sins pile up and trust the Lord is as merciful come judgement day as some say he is . . .'

Drawing breath while topping up our tea, Mum suddenly exclaimed, 'Wait on, lovey! I nearly forgot! She asked me to give a rose to a nice boy if he came asking for her. Margie's such a romancer – I guessed it was one of her jokes. I was buzzing around getting breakfast, so I says, "Stick it in a vase – and Robert's yer mother's brother!"'

'Margie says to me, "What is it you mean?"'

'I just had to laugh, because she doesn't always get our lingo, but she's as keen to learn as a parrot. "Bob's yer uncle! Job's right," I tell her. "On yer way, an' ava good day, love."'

'"I get you," she says. "Easy as the log is to fall off." And goes off laughing.'

Sitting back with a cuppa in her hands, Mum suddenly chuckled and met my gaze with twinkling eyes. 'My God! I'm as slow as a wet week! In the morning she's usually as solemn as a church owl, but she dances in laughing and joking this morning, and says, "Give this rose to a nice boy" – and I never woke!'

Mum's romantic heart was doubly delighted. Rising, she beckoned me close and pinned the rose through a button hole in my shirt, and then stepped back. 'Not bad at all!' she said. 'If I was free and twenty-one I'd take a ticket in you myself!' She threw her head back and laughed uproariously. 'My God, you should see yourself! You're as red as a rose!'

I could feel my face blushing as I stammered a self-conscious, 'Thanks', and hurried off in search of Doug.

My mate was sitting in the driver's seat reading a Penguin Ellery Queen mystery novel I had stowed in the glove-box for idle hours. 'Well, don't stand around like a stunned plover!' he said curtly, without looking up. 'Get in! You're a wee bit late – about three bloody hours late!'

I opened the passenger-side door, and the dog leapt from the back seat to the front and smooched his mate a welcome. Smiling vacantly, I patted Zulu. My wits were absent, my senses turned inward, cherishing love's afterglow. I looked in the back seat and mumbled, 'Is all the gear aboard?' and then gestured for Doug to release the bonnet so I could check the boot.

'Gees!' Doug said, exasperated. 'Have your ears gone on holidays with your brains? You never unloaded your gear! I've run Zulu and filled up with juice. Now bloody well wake up and get aboard!' Already he was revving the motor.

THE GYPSY COOK

We left Tambo behind, racing past the sprawling buildings and crossing the creek (top gear and cruising speed, seventy-two miles per hour) as we began the long haul to West End, Brisbane. The speedo slipped to sixty miles an hour as the Beetle mounted the long up slopes of the rolling downs, and ran out of figures at eighty on the down side.

With the VW wound out the Laughing Kiwi said truculently, 'If you had been at your post last night, soldier, we might have scored. Janice, the short dark one with the moppet top is a real sweetie. She was all my way – a good keen woman – but she wouldn't start without her mate. I reckon the lanky blonde was your type: good looker but quiet – she hardly said a word over dinner, except chatted to you about books. Call yourself a fisherman – you don't even know how to turn a nibble into a bite.'

Over the two years we had knocked about together I never found my mate to be naive or dishonest. Much as he loved the company of the fair sex, Doug never pretended to be a sound marriage prospect. The old saying, 'He likes the taste of honey but he won't keep the hive' fitted him like a favourite shoe. His line with the ladies was nothing if not straight: a few drinks and jokes to engage interest and camaraderie and, if the signals blinked amber, he'd say, 'Darlin', I'm here for a good time not a long time . . .'

Staring down the long road, he sang a few of his favourites to wear away monotony. 'Danny Boy', 'Comin' Through the Rye' and 'The Lemon Tree' infused my pleasant doze, before he called, 'Hey! Maybe Marion is a lemon – sunny on the outside and sour on the inside. When Marion opened the door, I asked for Janice, and I'll tell you Marion didn't appreciate me interrupting her beauty sleep.'

The Laughing Kiwi turned the radio up and drove hunched over the wheel, eyes glued to the lonely road, while I rested my head against the cushioned window and dozed, in wistful reverie of Marguerite.

17

THE MOTHER

A few miles on Doug swerved at seventy miles an hour to miss a suicidal wallaby. Centrifugal force slung me against his shoulder and then banged my head against the window. Close shaves weren't a rare occurrence with Doug at the wheel.

'Never mind your head, bro,' he said. 'Taking care of Pat's beautiful rose is the important thing.'

I picked up the cushion and settled back, smiling. I had told him I had picked the rose I was wearing for my mother, and now I imagined Mum's joyous hug as she received it. I recalled Marguerite asking frankly about Pat: her religious beliefs and nature, and my smile faded as I realised how similar their formative years were. There was not much room for leisure and laughter in either of their childhoods – the crucibles of World War I, family bereavements and the Great Depression for my working-class mother in Queensland, the loss of her mother and brother, and

World War II for Marguerite. But there had been books and flowers; and both had learnt to treasure small joys and count their blessings.

My mother recalled childhood visits to Brisbane. The War to End All Wars had depopulated Australia's cities and country of eager, trusting young men. As a young girl she had seen the returnees on the streets, crippled, mangled, amputated, blinded and often reduced to begging. They were shells of the robust confident heroes who had answered the Empire's call to enlist.

Many years later I came across a few survivors – proud old men earning their daily bread as station hands and rouseabouts. As a fifteen-year-old I had asked, 'Why does Andy camp away from the huts?' Long Charlie, shearing contractor and humane bush philosopher, had explained gently: 'He's shell-shocked, son. He camps away out of consideration for men who need their sleep. Sometimes he has bad dreams – he screams and yells . . . It's a subject he don't like to talk about. You'll remember that, young fella, won't you?'

The war's grim shadow had been darkened for my mother by the loss of her beloved father and elder sister, and the prolonged illness of her youngest brother – both siblings prey to rheumatic fever. Mum had shared the grief and burden of her mother – raising four kids on a country railway stationmistress's wages – before going west as a sheep-station governess to ride out the Great Depression. Enjoying station life and experience, she found the man of her life, and settled to raising seven kids on a bush battler's earnings through the latter Depression years and World War II, and the 1950s and 1960s.

THE MOTHER

My mother courageously travelled a hard road, often uphill against the odds – a road crowded by the out-of-work and out-of-luck battlers of her generation. Sustained by a calm nature, good humour, and 'faith in God and her good man', she complained little. Her favourite adages were 'Count your blessings' and 'Courage is the first virtue'. Now robust in middle years, she was thankful for a rewarding life.

I fingered delicately the petals of the rose pinned above my heart. It could only be a symbol of strong, passionate womanhood; Marguerite was so like my mother in many ways. The two women may have flowered in different cultures, but they were united in the strength of their Catholic faith, sharing a love of children, of music, of books and stories and (I smiled in pleasurable embarrassment) of *me*. One had loved passionately but briefly; one was bound by blood and responsibility – and life itself.

My mother, a gifted poet, loved to tell or hear a good story. She passed on stories she enhanced with the magic of drama, wit, charm and imagination, as yarn-spinners always have, from the shamans who honed their craft around prehistoric campfires to the bull artists of backyard barbies.

Fondly I imagined it would be fitting to gift my mother the rose and the story of Marguerite. I smiled: well, not quite all the story, for Pat's Catholic conscience might find the sensuality of the fulfilling chapter difficult. Besides, that part was for myself, alone; to be treasured among life's serendipitous blessings.

*

With a mile-post marching by every fifty seconds and the good times of Brisbane and the Gold Coast looming in imagination, Doug's cheerful chiacking returned. Halfway to Augathella he killed Johnny Cash to disturb me. 'Where the bloody hell were you, anyhow? I went door knocking at seven o'clock. Lucky I didn't run into a row – disturbing the good folk of Tambo saddling up on the wrong side of the blanket. It would have been your fault if I had copped a shiner.'

'Dunno. I was pretty drunk. I must have choked somewhere,' I replied with a small measure of truth. I turned my face into the cushion to relish a winner's grin: Doug's dedication to Aphrodite was relentless; and it was nearly always the Laughing Kiwi the goddess favoured.

'Choked alright!' Doug emphasised with a characteristic giggle. 'They thought you were dead and put a rose on your chest.'

'Ha! Actually, I picked it in the pub garden. I always take Mum flowers. She loves any flowers, but red roses are her favourites.'

'Mine, too,' said Doug.

I peered up at the sky. 'The weather looks okay. We'll give Augathella a miss, take the dirt track by Clara Creek and rev straight through to Roma, fill 'er up, grab a burger, and ring the old folks. We'll be home by seven o'clock.'

'Gees, mate!' Doug remonstrated in a pained voice, 'You're as handy as a candle on a windy night. Flowers are like women: you've got to kid to them, and look after them. That rose will be as withered as Adam's fig leaf by the time we hit Roma.'

'Yeah, I guess you're right. Pull over and I'll put it in the billy with some water.'

THE MOTHER

'Holy Dooley! A glorious rose presented to your mother in a bloody old black billy. Have a wee bit of respect! You just don't listen! Like I said, flowers are like the fair sex: they have to be flattered. Mate, I'll give you the *real* drum: we'll pull into the Augathella pub and have a couple of quiet beers – and get the girls there to hunt up a fancy vase for that beautiful rose.'

The Laughing Kiwi charmed a crystal vase out of the publican's wife. It set me back two quid.

With the hammer down all the way and the luck of the Irish, we rolled up the side lane of the old Queenslander at West End at seven-thirty.

My family tumbled down the stairs to welcome their long absent son and brother. There were warm hugs and kisses with my mother and sisters, shoulder hugs with my brothers, and a handshake and shoulder clasps with my dad, whose blue eyes moistened with emotion.

While the dog milled among people, jumping and yapping excited 'Hellos', I said to my younger brothers, 'Righto, fellas! Give us a hand to get this gear upstairs.' I turned back to get the rose to give to my mother and saw her dutifully hugging Doug. The Kiwi was embracing her with one arm and holding the vase aloft, out of Pat's sight, with the other. Disengaging and taking a step back, he formally presented the rose with two hands – as a priest with a chalice. 'Pat, I brought this rose all the way from Tambo – especially for you.'

She reacted with surprised delight. 'Doug! How thoughtful of you! My favourite rose; and presented in a lovely crystal vase. It's beautiful!' She hugged him again.

Realising I'd been trumped, I swallowed an exclamation of protest. My mate had taken my rose and my moment. On the morrow I would need to buy a bouquet to replace the rose. It would cost a couple of quid, but Mum would be doubly delighted, and Doug's puckish humour had sweetened her feelings towards him. We would soon share a laugh or two over the Kiwi's joke.

And Marguerite: I pictured her smiling quietly in the knowledge that her gift had completed its mission of love and laughter before it withered. Strangely, I felt she had foreseen its destiny, even as she stooped to pluck the rose.

18

THE CITY AND THE BUSH

Marguerite had taught me in a few hours how wonderfully warm and enlightening love can be. I was never to see her again, but I wrote of our experience and my feelings in my journal. The Laughing Kiwi stayed with us for a couple of days in December 1961 before flying home for Christmas, and I joined my family in our usual loving, laughing celebrations.

For some years a desire to become a writer had been niggling away at me, but I believed my fifth grade education would require a serious upgrade. I had saved enough to fund six months' study – funds which I intended to boost with regular pay nights as a preliminary boxer at Festival Hall.

Dad went to manage a small sheep station in the Blackall district, and Mum joined him in April, confident we would manage the household. Barry and Lyn were at work, and Carmel, Peter and Kevin at school. I bought books on maths, English and history, philosophy and layman's psychology, and holed up in my room to study. 'No beer and no girls – either would distract me,' I had declared to Mum. She smiled wisely. 'All work and no play makes Jack a dull boy', was one of her favourite axioms.

Being teetotal was easy, and the boxing began well, but asthma KO'ed me on my second week's training. It was a disappointment, for I had already begun to embrace the blokey competition and camaraderie, smell of sweat and liniment and the whack of leather. I kept plugging determinedly at the books, until a lung infection laid me up. My doctor recommended hospital. The Laughing Kiwi came in from shearing around Roma. He'd won a packet on Sharply, the 1961 Sydney Cup winner, and had some 'good oil' for the 1962 Cup. He suggested a drive to Sydney would benefit me more than a spell in hospital, and sure enough my health improved, while his 'good oil' galloped for the bookies' benefit.

Faye, a woman I had dated briefly in Hughenden, was in Sydney. We had kept in touch by letter. She had a 'fella', but made up a foursome one night for dinner with Doug and me, and with Birte Hansen, a Danish workmate. Faye was now a psychiatric nurse at Gladesville Mental Hospital, and Birte was an assistant nurse. A tall, blonde and bespectacled farmer's daughter, she had taken a two-year break from studying psychology to travel.

Back in Brisbane, I was engaged to join Richie Jack's team as a presser for ten shearers at Eulolo, Julia Creek, on the first Monday in July. Around 45,000 sheep would occupy five to six weeks, a task that would test me even when work fit, and I was underweight and run down. In the meantime, Dad engaged me as a station hand to help put through his shearing. I hoped that healthy outdoor work and Mum's cooking would build me up, and that the climate would clear up my asthma. Peter and Kevin would travel with me in the V-Dub for school holidays with Mum and Dad.

I was feeling a bit down: boxing had failed, and my sojourn of study was over. Although I had expanded my basic knowledge of maths and English and studied the books I had set myself, I realised I should have employed a tutor, for I had fallen well short of my objectives. A few days before I planned to depart, a blaring car horn drew me to the front door to see Faye and Birte, who had hitched from Sydney, tumble out of a red Mini-Morris. It was an unexpected and delightful vision.

Next morning I picked them up from their motel, and we made a fun tour of Brisbane city and suburbs, with a few stops for refreshments. Brisbane was Birte's first port of call hitching around Australia. Afterwards, she planned to travel to Townsville, Cairns, Darwin, Broome and down the west coast to Perth, where she was booked to board a ship to Europe in December. Hailing from a tiny frosty land, she said she 'would appreciate a trip west of the Great Divide', and asked for a lift to Blackall, from where she would hitch to the coast. With Birte in the front passenger seat and Peter, Kevin and Zulu in the rear, plus all our gear, the V-Dub was packed.

Birte was gentle, studious and humorous, and inevitably we became close on the 550-mile drive. She decided she would wait three weeks for me if I'd drive her to Townsville and she could find a job in Blackall in the meantime.

She picked up work as a housemaid at the Tattersalls Hotel and the Prince of Wales Hotel, while I did station chores, carted bales of wool to the railway at Yaraka on a three-ton truck, and Zulu and I mustered and penned up sheep and depopulated mobs of feral pigs. The shire paid a two-bob bounty for each pair of pig's ears, and Arthur Symes, the owner, offered ten quid for a rogue wild boar. 'The ferocious beast has only recently come onto my station,' he said, 'and he's killed three of my stud rams and maimed seven others.'

The tenner – over half a week's wages – would be handy, and the challenge was irresistible. *Ferocious, alright!* I followed outsize tracks along a bore drain. Unseen, the cunning quarry let me ride past, then charged from cover and hit my pony's hind legs. Tired old Flossie hadn't been above a trot in years, but she bolted so fast she nearly left me behind. It must have been a curious sight indeed: me trying to control Flossie, stay aboard and hang onto my rifle, while the tusker slashed at her fetlocks, and Zulu gave chase. The dog quickly lugged the tusker, and I wheeled Flossie and tied her to a tree. I ran back and shot the boar, then collared ten quid. Zulu got an extra chop.

The shearing cut-out. Birte and I left Blackall before sunrise, motored to Mackay, and loitered along the coast to Townsville, where we idled away a few days before Birte decided she wanted to see Hughenden. 'What a great idea!' I agreed. We took a room at the Grand Hotel, where Birte worked as a housemaid-cum-cook. I then got work 300 miles away in Eulolo, but would drive back on weekends to be with her when she wasn't doing the relief cooking on a Georgina River property. She finished up there a few days before Eulolo cut-out and waited for me in Julia Creek. 'The longest three days ever,' she said laughing. 'I would rather be snowed in for the winter in Norway.' Norway, it seemed, was the butt of many Danish jokes.

After Eulolo, I worked conveniently close to Hughenden while Birte again worked at the Grand. On a couple of weekends we camped at the beautiful Porcupine Gorge, which back then was a peaceful, secluded spot rather than the tourist destination it is today. We saw rodeos and shearing in full swing, and Birte insisted on standing front row to witness a savage bare-knuckle blue I refereed on the 'bull ring' at the rear of the Shamrock Hotel. 'Gladly I saw the fight, but I do not want to see again,' Birte declared at the end.

We grew more caring and dependent than we had intended, but nevertheless Birte and I avoided speaking of a permanent relationship. Birte longed for the homeland and family she loved; and I had no intention of marrying soon, despite the Laughing Kiwi's joking reference to us as 'the married couple', and asking me, 'How's the wife?'

I was still heeding the advice of my mother: *A man should wait till he's thirty to marry, when he should have accumulated enough property to support a family, and enough wisdom to understand a woman.* Or perhaps I was simply using that as an excuse. Either way, late in October Birte and I tearfully hugged farewell at Hughenden airport. She gave me her beloved copy of *The Song of the Red Ruby*, saying, 'Remember, Alan, that life is always beautiful.'

'But never in Norway,' I quipped, chuckling despite my damp eyes. She waved from the Vickers Viscount aeroplane, which was heading to Melbourne. From there she would take a memorable four-day transcontinental train journey to Perth. When she finally got back to Denmark she wrote of the train journey: 'I gladly did it, but I would not do it again.'

Doug and I completed Richie's run. We spent a few days with my family at West End, then drove to Sydney and flew to Auckland. After enjoying a few days with Doug's parents in Huntly, we shore with a Maori team for a week at Taihape. They were an affable crew, with cheerful girl rousies. But for Australians the tucker, which was mostly mutton and 'Maori bread' (damper) and spuds, was below par, and the accommodation was unhygienic. The wages and work hours were also uncivilised: nine hours a day plus Saturday mornings, with overtime at ordinary rates if required. A wool truck rolled up on our second Tuesday and we were on it, bound for the South Island, where the many Aussie shearers who migrated annually had installed Australian hours and pay rates. We shore at a couple of sheds, then caught the train to Invercargill. Doug returned to Huntly, while I travelled to Queenstown to ride the famous paddle wheeler to Glenorchy and hike a couple of trails. I was back in Huntly for Christmas.

Back in Queensland, we signed on at Booligar, near Dirranbandi, late in January. Doug returned to Hughenden for a crutching run, while I pressed around the south-west. Without Doug's cheerful presence I was assailed by a touch of the blues for months, and often thought of Birte.

In May I pulled the pin and retreated to Dingo Dell, a lovely peaceful location south of Prairie, to try writing poetry and short stories. Zulu and I camped in an old boundary rider's hut not far from Kooroorinya race track. Going into Hughenden for supplies, I ran into Doug who introduced me to his girlfriend, Joy, and her friend, Isabel, a pretty nursing sister. It was June 1963. I was instantly smitten. The four of us attended the Kooroorinya picnic race meeting ball, and within a few months I had forgotten my commitment to stay single until I was thirty, and Isabel and I were engaged.

Birte and I had continued to exchange fond letters, but when I told her I was committed she replied with four furious pages describing how I'd taken her for granted, and to write no more. I recalled lines from an old folk song:

When you go a-fishin', fish with a hook and line,
But when you go to get married, never look back behind.

Isabel and I married in Toowoomba in December 1964, and Doug and Joy the following year. They moved to the western districts of Victoria while Isabel and I settled in Longreach. Sadly, Joy died in a car accident a few years later.

Isabel got a job working at the Longreach local doctor's clinic and then moved to the Longreach hospital, while I scored a top wool-pressing run with UNGRA. We resolutely paid off a family home – an old Queenslander – within four years, in accordance with our plan to gain financial independence, and we settled into wearing down the edges of single independence to fit into the harmony of a loving marriage.

19

THE BROLGA AND SCREWJACK

On a wintry Monday morning my mate Thommo picked me up about two hours before sunrise. We had to hit the road early to arrive at the shearing shed for breakfast at a quarter to seven, to begin work at half past seven. It was 1967 and I had changed employers to GRAZCOS.

Thommo drove steadily, on the lookout for panicked roos that would bound out of the fading night at forty miles an hour, on a suicidal course to wreck headlights, grilles and radiators. He was a quiet and thoughtful man who read good fiction and popular history on lonely nights in the shearers' huts. His favourite hobby, however, was punting; and like all compulsive punters, regardless of their IQ, he believed that he could beat the bookies at their own game despite the mathematical certainty that the odds were stacked in the bookies' favour.

Also in the car was Tassie, a quality shearer from Tasmania, who could be relied on to deliver a flow of dry humour with an ever-ready grin.

After the usual exchanges relating weekend activities and family doings, Thommo commented, 'I wonder what appetisers Screwjack will rustle up for breakfast.'

'If it's anything like the last few weeks yer wouldn't feed it to drovers – or their dogs,' Tassie replied in his habitual drawl, as Thommo slowed and swerved to avoid several axle-busting potholes.

Like a lot of Australians of my generation, my natural inclination was to support the underdog. A few blokes were down on the Yugoslav cook because they saw him as an outsider, but most workers believed in a 'fair go' while on the job, regardless of a man's country of origin. I had come to see the babbler as a workmate, bound to his Yugoslavian heritage. He was a lonely outsider striving to fit in and be accepted as one of the boys. He signed on as Doug Grujac, but this had quickly morphed to 'Screwjack' on the nimble tongue of a shed wit. Only Thommo and I, dedicated readers, wondered what his role had been in the bloody racial and religious fighting during World War II, as the Nazis and communists fought for control of the Balkans.

'I wonder if he was a freedom fighter with Tito's partisans,' I had once suggested hopefully.

'I doubt he was old enough,' the cynical Thommo retorted. 'More than likely he was a smart-arse kid working the black market on both sides, a long way out of the firing line.'

Now and then Screwjack would waltz around the kitchen while whistling melodiously, before striking a pose and exclaiming, 'Hey, boys! I voz vunce the valtz king of London and gay Paree.' His capering drew little attention, bar an occasional: 'It's a pity you didn't learn to cook, Screwy, as well as valtzing and vistling.'

Camouflaged by the grey early dawn a doe and a joey roo bounded across the road from the shadow of a clump of gidyea trees. Thommo jammed on the brakes, and we braced ourselves against the dashboard. As the Holden regained speed I defended the cook. 'He's a pretty fair babbler when he's sober and on the job. And don't forget what he says: "You boys are lucky to haf the valtzing king to prepare your culinary delights." You should show some respect and appreciation, Thommo, for a brave and artistic freedom fighter.'

Facing a hard week's shearing to recoup his losses at the Barcaldine races, Thommo was in no mood for levity. 'Come off it, Presser,' he said, raising his voice above the revving motor and the rattle of gidyea stones striking the chassis. 'He's been with us for six weeks. Maybe he came up to scratch for a week, but he's been pissed every day since – and his tucker isn't fit for pigs! We've had three or four swarms, and you and me have stuck up for him to get a fair go. He hasn't got the message – and he's treated us like shit.'

'Yer right and yer wrong, Thommo,' Tassie drawled. 'I reckon Screwy is fit to cook for pigs! But not for shearers. I'd have speared him weeks ago if it wasn't for you blokes insisting we give 'im a go. I'll stick up for anyone, black, white or brindle if he'll avago, but if he won't do his job he's a goner!'

'Yeah, he's had more than a fair crack o' the whip,' I agreed. The recurring vision of Happy Jack, his body slumped in a chair, his head shattered, haunted my mind's eye as I continued, 'He said he was crook last week. Still, if he doesn't come up to scratch this week I won't be in his corner. But I won't speak to sack him – that's against my religion.'

'That's a bloody weak attitude, Presser –' Tassie began, but Thommo cut him short. 'No need to argue, fellas, because it won't come to a vote. Percy has got the message.'

Thommo was referring to Percy Taft, manager of GRAZCOS in the central west. Percy was an ex-gun shearer himself, and though he might have as many as 400 men employed in thirty sheds he remained hands-on with all his operations.

Thommo continued, 'I ran into The Brolga at the Barcaldine races. He cuts out a shed on Friday and starts with us next Monday. I reckon we can tolerate Screwy for one more week.'

'Bloody good show! Only one more week to poison us all,' Tassie declared.

'Bloody good show, alright! I've heard The Brolga is a champion, one of the best,' I commented.

'The creme de la creme, right off the top shelf,' Tassie concurred. 'I was with him a few years ago. They say he's a reformed alky now. He was a pretty good-humoured bloke for a babbler – when he was on the grog.'

Screwjack's tucker didn't improve over the next three days. He wandered his kitchen sucking stubbies when unobserved, and was full of boisterous bullshit at meal times. Thommo advised dryly, 'Suffer in silence, boys. Only a few more days and The Brolga will rescue us from Screwjack's abominable bowel-busters.'

On Thursday at ten to twelve, following my usual ritual, I swept the wool-room floor, put a new wool-pack in the Koertz wool press, washed my arms and face, collected the smoko tray and tea urn and headed for the kitchen a hundred yards away.

The six shearers followed a few minutes later tailed by the hurrying rousies, while the overseer and classer chatted in the rear. I was tucking into roast mutton, with spuds, pumpkin and cabbage nearly boiled to slurry, when Alec, the shed rep, swore vehemently and complained, 'This leg o' mutton looks as if it's been carved by a drunk with a tomahawk!'

'You can bet it has. Either that or the presser's dog has had a go at it,' Tassie said, as he investigated beneath two grimy tea towels which were covering the trays of prunes and melting jelly.

'Where the hell is Screwjack?' queried Thommo as he turned the tap on the urn. 'The bloody tea's not made. Have you seen him, Presser?'

'No! Maybe he's flaked in his room. He's been drinking like a fish since he blued with his missus. He says she drives him to drink. "Bloody vimmen! Drife a man to drink, alvays complaints, complaints . . ."'

'Come off it,' Thommo snapped, in no mood for my parody of the babbler's broken English. Thommo had shorn 102 sheep for the morning. He was looking for a good feed and twenty minutes' rest on his back on the shearing board to restore some energy for the afternoon's toil. Opening the back door, he lifted the lid off the tumbling-tommy. 'More stubby bottles than a Fourex bloody brewery,' he grumbled.

While we attempted to appease our appetites with hacked mutton and watery veggies, the overseer's angry remonstrance rang from the cook's bedroom. 'Screwjack, you lazy, drunken fool! Get up! The team wanted to sack you a week ago, but I saved your skin. This time you've done it. Get up and get to work or you're finished!'

We heard the cook groan and grumble, 'Sorry, boss. I gotta flu, really bad – sick as da dog. I get up soon and do da vashing ups.'

Realising that shearing would halt if he sacked the cook, the overseer changed his tone. 'Okay, Doug. See if you can carry on for the afternoon. Then pack up and get yourself to the doctor. I'll phone Percy and line up another cook for tomorrow.'

'Sorry, boss. I too crook. Da vashings up, I do. Den I go to town.'

Earwigging through the wall, Thommo commented, 'Too crook, be buggered! The lazy bastard has just run out of grog.'

Given the extraordinary circumstances the team automatically held a swarm on the verandah after dinner. Hickey, an ageing snagger, declared, 'We can't shear without a cook. It's against Union rules.'

It was said that Hickey had been a respected professional boxer in the hard times of the hungry 1930s, but the team knew him as an obnoxious old nark. He was fastidious in his habits, stood over timid rouseabouts, and kept to himself unless he had something critical to say. Yet, come Saturday he was a natty figure in Panama hat, sports coat and shiny shoes as he moved from pub to pub seeking the best odds from the SP bookies.

Tassie laughed. 'Hick, it don't matter who's the babbler, you wouldn't work in an iron lung.'

'Hick hails from the Big Rock Candy Mountains,' I put in. 'Where *they boiled in oil the inventor of toil, and hung the jerk that invented work.*'

A faint smile flicked across Thommo's face. 'I don't fancy the boiling in oil caper,' he said, 'but a swift death by hanging might be better than starving to death on Screwy's piss-weak stews. Botany Bay stews they used to call 'em in the old days: convict tucker! The lazy bastard dices a spud and a carrot and a shoulder of mutton in two gallons of water to feed twelve men.'

'Be fair to the wog!' Tassie objected. 'Yer forgot to mention the onion. There was one in last night's stew, alright. But it was such a bloody skinny stew Thommo could read the racing form guide through it.'

'Enough talk, you blokes! I'm calling the meeting to order,' Alec interrupted. 'The overseer isn't an AWU member, but it will simplify matters if he listens in.' There were no objections. 'He'll give us a few words to keep the show on the road this arvo – if we agree. As for me, I reckon we'd be better off making a quid than sitting on our arses waiting for a cook.' Alec was only in the game to save enough money to buy a small property for his family, and keen to get back to work.

'I vote we sit on our arses for the rest of the week,' said Fred, the seventeen-year-old picker-up. 'I don't mind getting paid for sitting on my khyber.'

Hick agreed. 'There will be no shearing till we get a cook.'

The overseer spoke quietly, as was his manner. 'I just phoned Percy; he'll send a cook early tomorrow if he can't get him today.' As shed overseer it was his business to get the wheels of industry turning with as little delay as possible. He waited for silence. 'He's a crackerjack cook. A chap they call The Brolga.'

'That's bloody good news,' Thommo said, recalling that on his word they had been prepared to suffer Screwjack's cooking until the weekend.

Who hadn't heard of The Brolga? Throughout the industry a handful of babblers were known far and wide for individual excellence in their profession. Vanity and touchiness invariably accompanied such excellence; thus the gun babblers were all prone to pull the pin at any word or deed they might interpret as insult. As their nicknames attested – 'The Bird of Passage', 'Wood Duck', 'Crack a Twig', 'The Bald Eagle', 'The Brolga' – they would all 'take wing' at the slightest criticism or provocation. Shearers were therefore careful not to upset a good babbler.

'A champion tucker musterer – drunk or sober,' Tassie drawled. 'I ran across him a few years ago. He was a mad punter. Used to go to town any Saturday he could score a lift, so we had to fend for ourselves for tucker. But he always came back that night – even if he had to get a taxi and fall out of it.'

'There's no need to worry about that,' the overseer assured us. 'They tell me he's been in Alcoholics Anonymous for years.'

'Too right,' young Fred concurred (Fred had been christened William, but had been dubbed 'Fred' after Fred Astaire because of the stylish steps he threw in while dancing along the board with a fleece). 'I done a shed with him last year. Drive yer mad he would, handin' yer papers about the evils o' drink an' bashin' yer ear with the Bible.'

The overseer looked around the expectant faces. 'Michael, the station manager, has volunteered to put smoko on this afternoon,' he said. 'He'll boil the billy and bring a sponge cake his wife made, and a few sandwiches. I'll help him to put dinner on tonight if Percy can't get a cook here on time.'

'We can't work without a cook; it's against Union rules,' Hickey declared.

'Hick,' I came in, 'the purpose of that rule is so that cooks will have a job. In this case Screwjack is already entitled to a day's pay, and we'll have a cook at sparrow's fart – so Union rules are covered. We'll be in breach of the Award if we bail up.'

The overseer held his hand up for quiet. 'I think that's settled to everyone's satisfaction, so let's go to work.'

Hick howled. 'We can't work without a cook.'

'Gees! Change the record!' Alec snapped. 'You don't shear enough to make a stew, anyway.'

The shearers and I – all contract men – voted to return to work for the afternoon on the understanding that a cook would be in place to prepare breakfast. Hickey objected but was overruled, while the rousies were despondently silent. They weren't entitled to a vote because their wages and tucker were guaranteed, work or no work.

Late in the afternoon Screwjack's beat-up blue wagon left a trail of dust as it headed for Longreach. Mike, the station manager, who looked like Chips Rafferty and had a sense of humour as dry as a drover's tonsils on a desolate stretch of the long paddock, brought the afternoon smoko to the shed and sat in to swap yarns with the boys. After draining his pannikin of sweet black tea he hauled his gangling frame off a wool bale and dead-panned: 'You fellows will be pleased to know that Screwjack was thoughtful enough to leave his recipe book behind so I'll know what to serve up to keep you happy.'

'Cripes!' Tassie exclaimed. 'More stewed tea and skinny stew.'

Mike produced a satisfying stew and roast mutton and veggies for tea. Percy and the cook hadn't turned up by breakfast time but, forewarned by an evening phone call, Mike was on the job at sparrow's fart to present a typical station breakfast of porridge, plus chops and gravy with eggs and bubble-n-squeak.

The shearing got underway at half past seven. At nine-thirty Michael delivered a tray of sandwiches and cake and an urn of tea. He usually loitered to yarn, but now he hurriedly downed a bite and a pannikin of tea and said, 'See yer later. I've got sheep to take away.'

'What about dinner, Cookie?' Tassie queried poker-faced. 'We can't work on empty bellies. You've got bloody good form, mate. I wouldn't say yer a gun babbler, but yer a sight better than Screwjack.'

'That may be so, Tassie,' Mike replied. 'But a man has to maintain some pride! I don't mind feeding pigs, but feeding shearers is beneath my dignity.'

At ten o'clock the team finished the smoko break. There was no sign of a cook, and another swarm talked things over while the overseer phoned GRAZCOS in Longreach.

'That's it!' Hickey declared. 'No cook, no work! We can't work without a cook!'

'It is *Friday*,' I said, making a point, and recited a verse I'd written:

> The week is past Friday at last, has come to ease the battle;
> We'll get aboard the big blue Ford, and into town we'll rattle.
> Our girls you bet will cease to fret, when they know we're handy.
> We'll douse the light and hold 'em tight, and let 'em know we're randy.

Alec said testily, 'I've heard it a dozen times, Presser. Dry up! With the missus a thousand miles away there's no use in me being randy. But count me in for a weekend in town. I don't fancy strutting around the mulga like a stunned bloody plover over the weekend.'

There were grins and nods of agreement all around: a long weekend was always welcome. All the shearers had families in Longreach they wanted to get home to, save Alec and Hickey. Hickey only worked for the necessities: a pub room and punting money.

We were on our feet and eager to pack up and hit the road when the overseer hung up the phone and called out, 'Hold it, you fellas! Jack, the owner, is flying The Brolga out. He'll be here by eleven to put dinner on.'

Reluctantly, the efficient little factory in the bush was soon back in swing to the regular thump of the Lister diesel and the rhythmic clack of the Koertz wool press, while the penner-upper bellowed, 'Pen-up, you bastards,' and 'Get over and speak up! Speak up!' to his kelpie, while the dog bounced over the backs of the woollies, barking commands.

Before eleven o'clock I saw the Cessna land and park near the huts. The pilot and passenger alighted. An hour later I stacked my last bale of the morning and glanced through the big open doorway of the wool room; an unfamiliar figure was approaching from the kitchen, walking as if having difficulty holding a straight line. This must be the famous Brolga, I mused – *but he can't be half-shot: this bloke is supposed to be a card-carrying member of Alcoholics Anonymous and a pain-in-yer-arse campaigner against the demon drink.*

I greeted The Brolga at the door with a handshake extended. 'G'day! They call me Presser, or Bluntie or Alan – take yer pick.'

The Brolga ignored the extended hand and said gruffly, 'I'm The Brolga.'

Dismissing the snub, I smiled and continued, 'Pleased to meetcha, Brolga. I saw yer coming in on the wing, and right away I reckoned you were some breed of bird.'

I chuckled at my joke, but The Brolga didn't join in. He stepped close and snapped into my face: 'A bloody smart alec! I won't cook for smart alecs.' He turned stiffly and headed for the door, calling, 'You're nothing but a bloody idiot.'

Joining me, the classer saw The Brolga halfway to the kitchen in full stride. 'Is that the new cook?' he inquired. I was glad he hadn't heard our exchange.

'What did he say?' he queried.

'He said his name is The Brolga – and then he shot through like a Bondi tram,' I replied.

We watched as The Brolga carried his port and swag to the plane and boarded.

'It looks like he wants to go back to town,' the classer observed.

Within seconds Jack approached. He opened the cabin door and appeared to be tongue-lashing The Brolga, ordering him to get out. The Brolga stayed put. Exasperated and angry but making the best of a losing hand, Jack strode swiftly to the shed to confer briefly with the shed overseer, before climbing into the pilot's seat. The Cessna taxied into the wind, hurried down the runway and took off.

The team shore till dinner time, and walked to the mess. 'There's no cook!' Hickey crowed. 'We can't work without a cook.'

'Yer can't help good luck,' Fred declared. There was no disagreement – and I didn't mention my altercation with The Brolga. As we ate the cold meat and salad meal The Brolga had knocked together, Tassie drawled, 'I told you fellas The Brolga was a champion babbler. This is the best feed to tantalise our taste buds in six weeks.'

We ate swiftly, showered, packed and put the hammer down on the road to town.

After dropping me at home, Thommo and Tassie had pulled into the Palace Hotel for a couple of beers. The Brolga came in and claimed them. He was 'flying high, powered by Johnnie Walker,' as Tassie told it. 'Cheerio, boys,' he gloated. 'I'm on the Viscount in the morning. I'll be backing winners at Eagle Farm tomorrow afternoon while you mugs are still here swatting flies. Then I'm off to Randwick for a few meetings; and then the Spring Carnival and the Melbourne Cup. Did I ever tell you I won ten grand the year Comic Court won the Cup? Quids in those days, my friends, not dollars. I was a professional punter for two years – *Professional Turf Advice*, my card read.'

'We need a cook, mate, not a professional bloody punter,' Tassie drawled.

The Brolga didn't heed. Confidentially close, he was giving Thommo the inside info for the next day's races at Randwick. Thommo said, 'I'll buy you a drink, Brolga, if you'll piss off!'

The drink came but The Brolga, convinced of his omniscience and feeling like a million dollars on Johnnie Walker Black Label, was already on his way, spouting expert advice along the bar.

'Let's leg it before he comes back,' Tassie said.

Thommo grinned. 'Do madmen go cooking for shearers or does cooking for shearers drive men mad?'

'I'm buggered if I know,' Tassie replied. 'I thought I'd seen 'em all; then we cop the waltzing wog trying to poison us for five or six weeks, and the famous Brolga, the champion of Alcoholics Anonymous, who turns up drunk as a skunk.'

Thommo said, 'I wonder who Promising Percy will find to grace our kitchen on Monday morning. I won't be surprised if Screwy's back – on the wagon, spick and span and full of bullshit.'

'You've gotta be jokin', Thommo,' said Tassie.

*

It was my turn to take my Peugeot ute to the shed. We pulled up at a quarter to seven on Monday morning. Observing smoke drifting from the kitchen chimney, Tassie said, 'Well, at least it's not Screwy; his old bomb isn't here.'

'Don't bet on it,' Thommo said. 'The powers that be would have ferried him out early yesterday – so he couldn't bring any grog.'

Screwjack, clad in ironed khaki trousers and shirt and clean white apron, presided over a spotless kitchen filled with the aroma of grilled chops and bacon. It was a challenging vision.

'Goot morning, boys! Thommo! I haf your favourite Uncle Toby's Oats and varm milk.'

Thommo didn't reply, and Screwjack continued, 'Doctor Murphy say, "Doug, you vera sick. No grog vile a you take da antibiotics." No flu now! Da King of Da Valtz is happy days! My goot vife vash and iron all my clothes.' He pirouetted while he whistled tunefully a few bars of a Strauss waltz, then wheeled and halted dramatically to address them all.

'Okay, boys! Listen now! Dis business is serious. Some bastard-shithead tell Percy I haf been on da grog. Is bullshit! Is der flu I haf. I find out I bust da bastard on da nose.' He swung a wild punch for emphasis, and we couldn't help grinning: the idea of Screwjack busting anything more substantial than a paper bag was ludicrous.

20

THE TIMES THEY ARE A'CHANGING

Through the early years of marriage I counted my blessings. Michelle, Helen and Jennifer arrived, all bright and healthy beloved babies. In September 1969, I took up an employment offer from the AWU to work as an Organiser, checking that Award conditions were being met and enlisting members. I was passionate about the work, but the job kept me on the road ten days of fourteen, the work followed me home, and the wage was ordinary. Three years later I quit, believing the job was putting stress on our marriage and that my immediate boss wasn't always doing the best for the Union's membership.

For a couple of years I returned to wool pressing before, in January 1974, we bought a dairy and milk delivery, paying top prices for cows. The cattle market soon collapsed and we couldn't afford wages, so I set about working seven days a week, rising at 1.30am, while Isabel returned to nursing. Marital relations deteriorated, we parted, and my wife began divorce proceedings. Our differences might have been many, but fortunately we agreed that the welfare of our girls was our mutual priority.

Six years on the authorities were putting pressure on fresh milkos to upgrade all equipment. The council demanded that I move the dairy outside the town limits and offered leasehold land. That would mean more debt and self-imposed slavery. I cashed the cattle, paid the debts and returned to the wool press to finance myself and assist with the support of our three daughters. With some free time available the writing bug nagged again. I published magazine pieces and poems, and laboured on a historical novel, which was never finished.

Now, away from the industry from the early 1970s to the early 1980s, I found the comforting familiar woolly smell and the pungent stench of the sheep yards remained, but the clickety song of the manual wool press had given way to the noisy, fume-spewing mini-motors powering hydraulic presses. The contract rate of pay was less but the work was considerably easier. I resented the cut in earnings but appreciated the lighter work load, as my muscles and joints were apt to remind me that I was getting a bit long in the tooth for really heavy yakka.

Ten years had seen dramatic social change. While most sheds were still fully unionised, I regretted that workers' solidarity and team spirit were dissipating. In the 1970s, Gough Whitlam, a progressive Prime Minister, had gone to bat for the workers, dramatically increasing wages. As a result there was a lot more ready cash about, and more shed workers had cars with radios and tapes blasting rock'n'roll at full volume on the roads Slim Dusty used to own.

'Mary J' had also come to stay. And in some sheds hostility ensued between young dope-heads, who claimed the right to blast the shearing board with head-banging sound, and older men who found the cacophony stressful and disrupting to their traditional quiet work rhythm.

Most young woolshed workers had abandoned the restful, money-saving weekends in the bush – spent reading, yarning, letter writing to Mum or a girlfriend, listening to the races, playing cards, fishing and hunting wild pigs and roos – for the thrills of getting boozed and stoned and the prospects of a weekend's shagging. The ACTU, led by Bob Hawke, had won 'equal pay for work of equal value', which was funding female independence, and the advent of the Pill had decreased the fear of casual pregnancy and increased sexual adventurism. Some men now 'shacked up' in town – short- or long-term – and it could be said that the hit song, 'I won't go hunting with you, Jake, but I'll go chasin' women', had become the theme song of a generation of young bush workers.

Although cars, telephones, electricity, diesel engines, radio and air-services had already modernised outback living, it wasn't until the 1970s that the work practice, mateship, ideals and traditional way of life of the shearing shed intrinsically changed. While female shearers and Kiwi shearers and their women rousies were yet to become a significant part of the industry, I realised mournfully that during my absence most of the elders had been marched to death or retirement by the remorseless drum of the old enemy – Father Time. The traditional shearers' campfire yarning, joking, chiacking and debating over politics, general news, women, sport and family were morphing from culture to folklore.

My heart was with some old-school shed overseers and shearers who battled to hang on to the quiet values of the way of life they felt comfortable with. Men of my own generation – hardy professional shearers of middle years who might come from as far away as Sydney's western suburbs or rural Victoria to support and educate a family – could find themselves shearing alongside tattooed, ear-ringed long-haired youths who often only shore enough to pay for their indulgences. Sex, drugs and rock'n'roll, backed by the relentless blast of tape recorders, was becoming the mantra of the shearing boards.

21

DICK, THE EAGLE AND RITZY

One Sunday after lunch I threw my swag and port on the back of my Tojo and called to pick up Lance. After the shearer hugged his wife and four kids we bought a couple of cartons of Fourex longnecks at the Midlander Hotel, crossed the Thomson River and headed north along the rutted river-road. Billowing dust, the ute crossed Mitchell grass downs and stony gidyea ridges, and dipped through bone-dry gullies which would become brown torrents rushing towards the river when the summer rains came – if they came.

Lance was a conservative, dry-witted bloke of about thirty. Six foot, lean and fair, he was as straight as a boree telephone pole in figure and character; and a fast, clean shearer. If not a big gun he was certainly quick and consistent enough to keep the would-be's and tearaways honest from Monday to Friday.

We passed through Muttaburra and followed the Hughenden road until we turned off at the station mail box. Five miles in we passed the homestead, a friendly old low-level Queenslander sitting among shady fig trees, bauhinias, poincianas, jacarandas and citrus trees.

A wide, flowing bore drain, its banks bound with green wild couch-grass shaded by clumps of prickly acacias, emerged from the house yard and flowed for some 400 yards before it swung around the shearing shed, watered the sheep yards and holding paddocks, and followed the fall of the land through the property.

'Great place to lay down a few beds,' commented Lance, a keen gardener. 'Beautiful pure bore water up this way, sweet to drink, and veggies love it. Longreach bore water is too acidic. Use it for a year or two and your soil is stuffed.'

'Yeah,' I agreed. 'Not much of a drink, either – unless yer as dry as a wooden god. Tastes like pig's piss. I guess that's why they call it *boar* water.'

Lance ignored the worn-out pun. 'Speaking of grunters,' he said, and pointed to a big white sow with a young litter lying under an acacia. Close by half a dozen muscovy ducks were paddling while a couple of tail-swishing stock horses, a bunch of red roos, three emus, a Jersey cow and calf and a couple of nanny goats with kids watched their passing. 'Strike a light!' I exclaimed. 'If it's not Animal Farm or the Western Plains Zoo.'

'Go for Animal Farm,' Lance urged, pointing to a huge white boar standing in the middle of the bore drain. 'If that's not Snowball I'm the town drunk!'

Smoke blowing from the kitchen chimney and a brown Ford wagon outside the overseer's quarters were the only signs of occupation. 'Dick Buchanan is boss-of-the-board,' Lance said. 'Looks like he brought the cook; there's a few Hughenden blokes coming to make up the team.'

We unloaded our gear and made up our shearers' stretchers in a room on the eastern side of the huts – away from the afternoon November sun.

I had met Dick twenty-odd years earlier when I'd been pressing for the United Grazier's Longreach office, and Dick had been a cheerful stud rousie (learner classer/overseer). From the sunlit plains of Longreach he had gone to serve in the jungles of Vietnam. He returned a changed person, who spoke sparingly of his war experiences. After that I didn't meet him for years, during which time he became a periodic heavy drinker and his marriage broke down. At times he was morose, but otherwise still witty and natural officer material. I saw him as a survivor from an earlier more romantic era, a latter-day Breaker Morant, who still now and then earned his daily bread as a contract musterer and horse-breaker.

Dick was sipping over-proof rum diluted with water while preparing sign-on contracts when I walked in. After the mandatory 'G'days' and a handshake Dick offered me a drink. I avoided rum and it was too early in the day for beer, but we briefly caught up on old times before the overseer suggested he had immediate duties by picking up a pen.

Before leaving I had one important question. 'Who's the babbler?' I asked.

His pen poised, Dick looked serious. He liked to quote verse and proverbs, and snippets of wisdom and humour; now he drew on shearers' folklore:

The greasy cook had a sore-eyed look;
He was covered in dust and ashes.
He stuffed our holes with his doughs and rolls
And he'd poison a dog with his hashes.

'A bait-layer!' I recoiled, suddenly feeling dull. The prospect of three weeks in the mulga having my digestive system ruined was discouraging. 'Yer gotta be joking, Dick.'

Dick replied with mock severity, 'Bait-layers don't last long with me. You ought to know that!' He broke into a grin. 'Cheer up! They call him The Bald Eagle. You must have heard of him? He's an old digger – loves a bet. Affable old bugger, when he's not whingeing or trying to stand over some rousie. He's been with me since July. Bloody trimmer of a cook; I've never seen better.'

I had found chucking a shovel full of complimentary bullshit into the kitchen usually made a fair start with cooks, who often felt deserted and unappreciated, so I pushed open the gauze door and called breezily, 'G'day Eagle! Dick says yer still turning out some of the best tucker in the back country.' The Eagle looked to have fined down a bit over twenty years, but he was over six foot and bulky, and clad in a chef's cap and whites. Turning away from the meat mincer he had attached to the work table, he sluiced his hands in the sink and dried them before replying, 'Dick wouldn't know cordon bleu from an army kitchen!'

'I haven't run across you since the rabbits were bad,' I said. 'How are you, anyhow?'

'How am I? How the bloody hell would I be – tucker-mustering for a mob of belly-achin' greasies at my time of life?' The grin on his face belied his aggrieved tone; maybe he had become affable.

'Who the bloody hell are you, anyway?' he demanded, reaching for a handshake.

'I pressed for Alex Meekin around Quilpie first half of 1960. We did a shed or two together. You were the babbler – a bloody good'un, too. We haven't crossed roads since then as far as I recollect.'

'Gotcha now!' the Eagle confirmed. 'Green Volkswagen, blue dog . . . a mate of mine. What was his name – Zulu?'

'You've got a memory like Sherlock, Eagle. A wise hound, that Zulu: first thing he did when he got to a shed was pal up with the cook.' Like a good mate or a lost love, I still missed him.

The Eagle turned back to a big mixing bowl containing diced mutton, kidneys and onions. 'If you're set on chin wagging – wind that handle and make yerself useful,' he said. He added a pinch of curry, a few spices and breadcrumbs, then attached a cylinder of sausage skin to the spout of the mincer. I wound the handle while the cook, flicking his wrist like a table-tennis champion, laid out strings of tasty snags.

We talked over old times and acquaintances, but never mentioned his fight with Jimmy G over twenty years earlier. For years the famous fight in varying versions had provided ready entertainment when shearers spun yarns around pubs and campfires. I was at first surprised that blokes I didn't know from a bar of soap swore they had witnessed the famous affray: 'Fair dinkum, mate – I was there! You've never seen anything like it. What a bloody turnout! Blood was spattered around the kitchen like a killing block. The

Bald Eagle was on his way to hospital and Jimmy dancing around like a bantam rooster challenging all comers.'

It didn't matter that names and locations often changed as the story evolved, for in any society – at least since David KO'ed Goliath – the little bloke knocking over the big bully has been a popular angle. However, The Eagle's order to 'Shut the door! Were you raised in a tent?' and the reply, 'Maybe it was Jimmy Sharman's tent,' remained the constant punch line.

Like me, The Eagle had returned to the wool industry after a long absence. 'The old Eagle has roosted in a lot of places,' he said, as he spun out the snags. 'I've fed school kids, soldiers, nurses, even nuns; truck-stops, cafes, pubs. You name it. The Eagle was top chef at a Chinese restaurant. And I had my own restaurant for a couple o' years – in Fitzroy.'

'What happened?' I asked, still winding the mincer. 'You must have made a quid. Did yer punt it up?'

'No chance,' The Eagle responded with a note of despondency. 'Woman I was with for five years – Sheila – talk about style: a double for Dolly Dyer! She was alright – managed the place, and she wasn't afraid to get her hands dirty. But then, out of the blue, she employs a new head waiter, a smoothie, a Spaniard.'

He laughed bitterly and fed the mincer. 'I paid him off and told the bastard I'd kill him if he came near the place. He spun me a line and apologised. He said he knew he had done the wrong thing.

'"You are her big man," he says. "Please forgive her – for woman is weak and man is strong. God will punish me for my sins. Boss, you are good *hombre, numero uno.*"

He said I wouldn't see him again – and I didn't. She made up lovey-doveys with me that night, and the next day she collared the loot from the cheque account and they hit the toe together. I tried to hold on, but I hit the grog . . . No fool like an old fool, as they say.'

'Sounds like an episode out o' *Fawlty Towers*,' I said, laughing. 'Did he go by the moniker of Manuel and come from Barcelona by any chance?'

The big man stopped feeding the mincer. 'It's no laughing matter – nothing to joke about at all.'

Often at the start of a shed the first meal was a hit-and-miss affair while the cook organised himself and his kitchen. Not so with The Eagle. His pride made a point of getting to the shed early to prepare to hit his straps with the first meal.

More vehicles rolled up late in the afternoon, and The Eagle had a full team for tea. Enticing smells emanating from the kitchen drifted along the verandah and through the huts as we unpacked, made up beds and had a beer. We were on our feet and eagerly following our noses before the bell ceased clanging.

A commanding figure as we entered the kitchen, The Eagle let everyone know we were welcome in *his* domain, but there were rules: 'Line up boys, no rush; there are the plates, knives and forks, always in the same spot. Tidiness is my middle name. Mutton and veggie loop-the-loop – deliciously flavoured with The Eagle's oriental spices – and Spanish mornay for entree. Two cutlets a man, plenty of snags and enough mashed spuds to feed the Irish brigade, plus peas and pumpkin. Make your own toast, and there's tea – and coffee for the connoisseurs. There's a plate scraper near the scrap bin: use it! Left-over soup, and cocoa till nine o'clock, but clean up your own mess, and be out of the kitchen quick smart: The Eagle has to rise while you jokers have got your hands on your cocks and are dreaming of sweet Adeline, so keep it quiet!'

Chances were we could look to a cheerful shearing: the northern sheep were free cutting and combing for good tallies; Dick was a fair and square boss-of-the-board who often raised a laugh with dry, witty repartee; the babbler was top shelf, and there were no card-carrying cranks or snags in the team. Plus there was a respectful teenage jackaroo (the boys dubbed him 'Silent') who was learning how to run a station; and the cocky, Jeff, and his ringer, Kelpie, were hands-on blokes who looked forward to a yarn and a laugh at smoko time.

There was the usual turn-up of entertaining individuals you'd find in a shearing team. Ralph White demanded a lot of space and attention. Dubbed Ho Chi Minh because of his resemblance to the Vietnamese revolutionary (he was also known as the 'Chinaman'), he was the chiacking, obnoxious, sarcastic, entertaining epitome of the old shearers' adage 'He don't shear many – but he's bloody good company in the huts!' A bantam rooster, often crowing with little man's aggro, he could, in fact, shear pretty well when he cut back on a variety of illegal substances and booze – for a man who might weigh nine stone in a wringing-wet army overcoat.

The shed overseer loved his rum but wouldn't allow marijuana. He had quietly fronted Ho Chi and his roommate, Wardy, and a wool roller before the sign-on: 'No dope around the huts or the shed. Get that! One whiff and you're down the track!'

'I'll do what I like in my time,' Ho Chi replied resentfully.

'Try it on, *cobber*,' Dick said as he walked away. Consequently, the smokers regularly took long strolls along the bush roads after tea each night.

Ritzy was elected Union rep. He was fair-haired, stood about five foot nine and was pushing fifty, but his slender build and quickness of action belied his age. He had been a noted gun shearer in his early days, but no longer felt the urge to prove himself by taking the lead. 'Only a fool lives to work,' he would declare. 'I work to live.'

Ritzy and I had worked together on and off from 1958 till 1963, when we travelled with a Hughenden team to a station near Winton. A Winton gun had joined the team as the sixth shearer. He was fast alright, and he let the team know it; and he got under the Hughenden gun's skin from the word go.

Ritzy had taken him on, and they had gone at it sheep for sheep. After four days of unrelenting combat, when Ritzy was in the lead by thirty-six sheep, the Winton man had said he had the flu and pulled out. Ritzy then dropped back and shore along comfortably a few sheep a day behind the Laughing Kiwi. The Kiwi reckoned Ritzy had his measure any time he chose to draw the whip, but he was a fearless stirrer: 'C'mon, Ritzy. Have a crack at a Kiwi – and eat sheep shit from go to whoa!'

Ritzy laughed, and shore along behind the Kiwi until cut-out day came. Their tallies were dead even, setting the scene for a classic shearers' duel. Ritzy went into top gear and McMillan went with him. The Kiwi gun pushed the Hughenden man to his limit. When the shed had cut-out at two o'clock Ritzy was the ringer – by one sheep. It was a cool winter's day, yet when he pulled his grey flannel singlet over his wiry shoulders and wrung it out, a stream of salty sweat puddled the shearing board around his bag boots.

Ritzy was a swift mover, deft and organised; he had unloaded his bog-eye and packed his tool bag before Doug pushed his last shornie, a hard-cutting cobbler – the toughest sheep in the pen – down the shoot. Laughing, Ritzy patted the weary Kiwi on the shoulder. 'You keep studying my style, son,' he said, 'and you'll soon learn how to shear like an Aussie.'

With the shed in order, Dick rang the bell and the team pulled into gear. Perhaps Ritzy could have taken the lead, but he settled comfortably about fifteen a day behind Lance, who was close to the 200 stroke.

On the second morning Wardy installed his tape player on a disused piece-picker's table under a window. He revved it to full blast, caught a sheep and pulled into gear. The discordant din drowned the chatter of the workers on the wool-table and buzz of the hand-pieces. Ritzy shore on, rhythmically placing his blows with the skill that only comes with dedication and long practice, shaping his sheep with artistic energy until he rolled it, a sculptured masterpiece resembling animated white marble, down the shoot.

Then he strode to the tape-deck, extracted the tape and pulled Wardy's machinery out of gear. 'Listen, mate!' he addressed the druggie, who was half a head higher, but looked like a witless drone with a half-shorn sheep between his legs and his silent bog-eye in one hand: 'My advice is that you file the corners off this tape and grease it with Vaseline, because if I hear it again I'll shove it fair up your arse!' The recorder was heard no more on the board, the rousies resumed their chatter and shearing progressed in relative tranquillity.

22

JOE BLAKES AND A RAMBLING POMMY

Later that morning as we yarned over smoko, Ritzy drew attention to a big snake's track outside the wool room. 'As a matter of fact,' he said gravely, 'I've recommended to the Union organiser time and again that the Union should ensure that wool pressers are paid "snake money". Our presser here –' he went on amiably '– might recall that twenty years ago at Barenya I warned him to be careful. In those days I had only seen one presser die of snakebite, but since then I've seen two more pressers delivered to the arms of Jesus.'

Because Ritzy was a noted yarn-spinner the boys tuned in. Cyril, a big Yorkshire lad, had been in the outback only a few weeks, but already he'd been dubbed the Rambling Pommy and been fed enough dire warnings of deadly taipans, death adders, Downs tigers and strangling pythons to make him tremble whenever the word 'snake' came up.

Jeff hid his grin behind a large hand and a pannikin of tea, before he objected: 'No snake money here, Presser! We work by the Award. Any demand for snake money and you can roll your swag.'

'It's not bad odds,' Ritzy went on, as poker-faced as a judge donning the black cap. 'Thirty-odd years in the game and I've only seen three blokes killed by snakebite – but they were all pressers. A lot of people think a snake strikes like a whiplash and pulls back, and then it's all over, drover! That's all bullshit! I can tell you a big Downs tiger, when he strikes, hangs on like a pig-dog swinging on a grunter's lug. And I swear, boys – on the Bible – the dreadful memory of Mickey's fatal screams and contortions still gives me nightmares to this day.'

Apparently stricken to silence by the horror of the memory, he went to the tea urn and refilled his pannikin. He added four teaspoons of sugar to the brew – which he'd sip to maintain energy over the next two-hour run of shearing – while the Rambling Pommy turned pale and a variety of voices urged Ritzy to: 'Get on with it!'

'Kelpie' Tom, a lanky ringer-cum-drover from Charters Towers, west of Townsville, was stretched out on the wool bales rolling a smoke. 'You take yer time, Ritzy,' he drawled. 'When yer start shearing I'll be obliged to leave this here air mattress to straddle a saddle an' fetch a mob o' hoggets for you hungry bastards to undress. You take all the time yer need, comrade. It ain't often these young fellas get the opportunity to hear a real expert discuss herpetology.'

Ho Chi Minh, compelled by his little man's complex to grab attention, declared loudly, 'Herpetology! That's too big a word for shearers, Kelpie. For you blokes who don't know, herpetology is the study of snakes and reptiles and amphibians.'

'Is that so, Chinaman? I'll remember that. I only threw it in for flavour, an' I was wonderin' what it meant meself till you elucidated.'

Gaining the respect of silence and attention, Ritzy resumed. 'We were having smoko that awful day. The presser was catching a bit behind, so he walked into the bin and began gathering a big armful of wool. We were sitting, just like we are now, when we heard this *bloody awful scream* – like a dingo bitch giving birth to a rabbit trap – and the presser burst out of the bin going backwards with a seven-foot Downs tiger swinging off his Adam's apple. We knew it was a death grip! And everyone was paralysed except poor Mickey. He leapt and rolled and shrieked like the voodoo dancers you see in the movies; he went into awful contortions, like that sheila in the movie *The Exorcist* – only this was fair dinkum. He backflipped and fell flat on his back with the big Downs tiger on top of him. It took two of the boys, with their feet braced against poor Mickey's shoulders, to grab the snake and pull and break the death grip. The snake broke free and went after more victims. We stood on wool bales and watched him thrashing around the wool room like an electric cable gone mad until the boss blew him to pieces with a shotgun.'

'I do recollect yer advice, Ritzy,' I said, stoically. 'Since you warned me about taipans and death adders camping in the wool bins I've bloody near frozen whenever I've been arming wool and a mouse has run up under my singlet.'

The picker-up, sixteen-year-old 'Fritzy', was a greenhorn rousie from Brisbane. He and the Rambling Pommy were the only two gullible enough to swallow Ritzy's yarn. Grinning, the team headed back to the shearing board, while Kelpie donned a mask of solemnity and insisted on shaking hands with me. 'If that old Joe Blake fangs you, friend, we won't meet this side of Judgement Day – so I'll bid farewell.'

The giant snake's track and the ground-level wool room, which gave the reptile easy access, combined to make me wary. I quietly cut and trimmed a springy snake-stick from a branch and kept it handy.

On Monday morning Ritzy and Ho Chi were seated in the wool room eating smoko when Ritzy mentioned snakes, as he often did, 'to keep you sharp and alert, Presser, old mate – because I doubt my nerves are strong enough to see another man die in the awful convulsions of snake-bite. This "Joe Blake" could be a harmless carpet snake or a young black-headed python; but more than likely this track is the call sign of a deadly Downs tiger, a seven-foot man-killer. More than likely he camps in the wool bins, old mate.'

Ho Chi promptly objected. 'No way! Judging by his track this snake is too big to be a Downs tiger; they don't grow over five or five-and-a-half feet!'

I said, 'Around Hughenden and in the north-west, where I worked with Ritzy, they call a king brown or mulga snake a Downs tiger; but around Longreach it seems a Collett's snake is often dubbed a Downs tiger. Take yer pick!'

I admired the beauty of snakes. I always found their sinuous movement fascinating, and I would watch their intricate patterns of colours camouflage their swift disappearance as they mingled with multicoloured soils and grasses. So it was with regret that I killed the snake outside the shed later that morning with three quick strikes of the snake-stick. 'Sorry, brother,' I said aloud in regretful justification, 'but it might have been you or me – or the station kids, or the dogs.'

Recalling Ritzy's well-earned reputation as a joker, I raided the first-aid kit for sticking plaster, taped the dripping fangs and put the coiled reptile in a corn bag before sauntering along the shearing board to stand in Ritzy's catching pen. 'More wool, more wool you presser-starvers,' I shouted, as Ritzy strode past me to catch a sheep. 'Yer kids will miss out on Christmas if you presser-starvers don't fill the comb and push a bit harder.'

Ritzy ignored me, then caught a ewe and dragged her to his stand. As he bent and positioned the ewe to shear the belly-wool I slipped up behind him, slung the gear cord over my own shoulder and dangled the snake in its place. The picker-up, rising from the next stand with a fleece in his arms, saw the snake and stumbled bum first into a locks-butt, while Ritzy reached back blindly for the familiar cord and seized the snake. Calmly turning his head without straightening his back, he glanced at the dangling killer: 'Just as I predicted – a bloody great Downs tiger. Now pull me into gear, Presser – and *piss off.*'

23

THE LORD OF THE FLIES AND FLYING EMUS

Some thirty yards from the kitchen, beside the walking pad to the shearing shed, a 2000-gallon tank perched on an eight-foot-high stand. At shearing time it was regularly pumped full of sweet artesian bore water for the workers' drinking water, kitchen and showers. The tank sometimes overflowed, forming a muddy puddle beneath, offering the great white boar a shadier midday wallow than the piebald shadows cast by the prickly acacias along the bore drain.

The pig was wont to grunt courteous greetings to sweating toilers as they passed to and fro on their dinner excursions. Some replied politely with 'G'day, you lucky bastard,' while others, feeling betrayed by a fate that kept a pig in porcine luxury while they sweated for their daily bread, gave the boar some obscene advice. Ho Chi – when he was in the mood – bowed his knee, doffed his greasy tweed cap and requested, 'Permission to pass, Lord of the Flies?'

On the Friday morning Ho Chi surfaced in a foul mood, and the boys surmised that 'the Chinaman' was out of dope. Over dinner he stridently demanded the pig's removal. 'It's not hygienic to have a pig wallowing in pig shit this close to the huts.' Getting no response he continued his tirade: 'It's against Union rules and it breaches the health act. I can name twenty-seven diseases that are common to Homo sapiens and pigs.'

'Well, name just five or shut your trap,' Lance suggested.

It was doubtful that Ho Chi could name any diseases common to man and porker, but ignoring Lance, he soldiered on. 'If some idiot didn't encourage the pig by feeding him under the tank he wouldn't be there – he'd be in his f—ing place, in the bore drain.'

The Eagle, who had palled up with Snowball and served him buckets of scraps each day, broke in with a bellow. 'Lay off old Whitey, you runt. Pigs are clean. They're not like a lot of people: they don't shit in their nests! I'd rather cook for pigs than a lot of shearers – like big-mouthed runts.'

In the knowledge that good babblers were almost as rare as bunyips, while cranky snaggers fell out of trees like ripe mangoes, the boss intervened. 'Whoa, Eagle! Hold your horses. Snowball will be okay. Ho Chi and the rep and myself will make an inspection as we return to the shed. I'll see the porker gets a fair crack o' the whip.'

Most of the team watched as Dick fetched a seven-foot kangaroo-hide stock whip he'd plaited himself and went into action a few yards from where the white boar lay at ease. After warming up with a stockman's crack and a few round-the-head 'window-rattlers', he announced each combination before he swung the 'dreaded lash': a couple of 'Sydney flashes' to begin, then left- and right-handed 'tom thumbs' went off like a string of crackers, followed by a battery of 'Waterloo cannons', a 'drum roll with thunder claps' and 'one for the ladies'. As the echoes of the exhibition bounced off the shed and huts and died away, Snowball, wondering what all the fuss was about, rose indifferently and wandered a few steps in the direction of the bore drain.

Dick tendered the whip to Ho Chi, and said sardonically, 'Well, old chap, Snowball is on his feet and pointed in the right direction. I call that mission accomplished. Of course, you're welcome to urge him on – but don't apply the lash: that might bugger the cracker – and it would stir the cook . . . And that I *won't* tolerate.'

The little shearer was aware of the limits of his whip-cracking ability. Declining the opportunity to make a fool of himself, he marched for the shed, as straight-backed and recalcitrant as Napoleon in defeat.

Snowball grunted his pleasure and turned back on hearing The Eagle yodelling 'Piggy, pig, piggy, pig . . .' as he tipped half a bucket of scraps into an old face-dish he had placed by the pig wallow.

The next morning the team was chatting and chiacking over smoko, when three tame emus appeared pecking after grasshoppers and grubs along the green verge of the bore drain.

'Can emus fly?' the Rambling Pommy asked innocently, in his broad Yorkshire accent.

'They can. Indeed they can!' I declared solemnly. 'But it's a mighty rare event; one I've only been privileged to witness once or twice.'

Experienced ringers regularly turned up at the shed at smoko time. Kelpie Tom, squatting on the flagstone floor stockman style with a pannikin of tea and a slab of brownie, contradicted me. 'Not so, Presser! Up in my country, in the Basalt, nor-west o' the Towers – where the dingoes gather in packs of a thousand or more for a moon-light howlin' shebang – you'll often see emus on the wing.'

'But they don't gain much elevation or distance,' I argued, thinking the yarn might grow too ludicrous if left to Kelpie's unbridled imagination – even for a Pommy's digestion. 'In fact, by my observation they're short, fast, flat flyers, like giant quail.'

'Right and wrong, Presser! Let me put you straight. Up in the Basalt country the chicks learn to fly or perish; and I've seen a grown bird, pursued by a mob of howlin', blood-thirsty dingoes, fly a mile and perch thirty feet up in a bloodwood tree.'

Jeff had entered and poured a mug of tea, and seated himself on the bale the ringer was squatting against while Kelpie enlarged on the flying ability of emus. 'Now Jeff, the boss of the show, is a bloke who stretches the truth now and then – which does not qualify as lying. So I believe him when he tells me that he was up in a helicopter with a contract musterer, well above the treetops, when they viewed a daddy emu with twelve chicks flying south in a V formation at a thousand feet, pointing like a mob of wood-ducks. Of course, yer know it's always the cock bird that looks after the emu chicks.'

Jeff trickled hot tea down the back of Kelpie's shirt. The lanky stockman squirmed. 'I didn't hear you announce yerself, Jeff. But what I say is ridgy-didge, isn't it?'

'If you say so, Kelpie, if you say so.'

Fascinated, the Rambling Pommy asked, 'How fast do they have to run to take off?'

Ho Chi chipped in authoritatively. 'An emu's wings in flight have been scientifically recorded at over a thousand beats a minute — faster than a humming-bird's. The ratio of wing span to body weight means an emu has got to have at least 980 wing beats per second and run at a ground speed of fifty miles an hour to lift off.'

Kelpie had eased on to the floor, and now lay at full stretch on his back. Smoke and laconic speech issued from beneath his work-worn Akubra, which rested on his nose and shaded his eyes.

'You are right but you are wrong, Chinaman,' he returned dismissively. 'Obviously you have never seen an emu attain flight. Any ornithologist worth his feathers knows that the emu, when alarmed, runs a few speedy steps to wind-up revs, pirouettes a la Dame Margot Fonteyn, then goes straight up like a helicopter.'

'Do they attack people?' the Pommy queried nervously.

'They do not,' Kelpie stated flatly. 'But keep an eye peeled for low-flying emus. Like an ocean-going liner they are not agile at changing direction. If you hear them coming the best shot is to drop as flat as a lizard.'

'Hear them coming?' asked the puzzled Pommy.

'My bloody oath! The wing beats of half a dozen emus howl like a Stuka dive bomber coming in for the kill.'

Ho Chi butted in assertively, 'You're wrong about emus not attacking people. I've seen a pet male get very aggro, drumming and scratching and running right at me. If I hadn't thrown my hat over his head and legged it I'd have lost an eye.'

'That's just it, Chinaman! If you had the guts to stare that bird down you would have called his bluff. In fact the only case I can recall of an emu hurting a human occurred up in the Basalt a few years back – and that was not the bird's fault! A small boy, a city kid, was out riding on his own. The pony was an old hand with emus on the wing, and when he heard the terrifying howl of an approaching emu flight he dropped to his knees and lowered his head for safety. But the little fellow stood tall on the saddle to take a gander and see what the noise was about. A split second later he was run clean through by the bird's beak and carried away, swingin' an' squealin' like a porker on a Bengal lancer's point.'

The listeners twisted their smiles to frowns and choked their laughter, while Cyril sympathised: 'Holy smoke! That's bloody horrible. The poor little chap!'

'Don't make yourself too comfortable, Kelpie,' Jeff interjected. 'We've still got a thousand ewes and lambs to draft. Fact is, if you were as keen on toil as you are on talk we'd have half the mob through.'

The ringer's rangy frame rose in reluctant sections; his two red kelpie dogs, snoozing obediently in the corner, sensed his move and were instantly on four paws, ears pricked and eyes shining.

'Another point worth noting, Rambler,' Kelpie mentioned casually as he exited, 'if you take a photo of an emu on the wing you're as rich as Rockefeller. Photos of flying emus are rare as big-hearted bank managers.'

The next morning an excited Rambling Pommy brought his Canon 35mm camera to the shed. After swallowing a quick smoko he commandeered a reluctant presser and an eager Fritzy to chase emus, and proudly led his team on a *Boy's Own* magazine adventure to photograph flying emus.

While the watchers in the wool room doubled up with laughter, the amazed tame emus ran and circled and dodged, always a step ahead of our yelling pursuit.

I came back grinning sheepishly, while Cyril and Fritzy, despite failure, were winded but undaunted. 'Did you see that?' Fritzy shouted. 'They had their little wings out; we nearly had them flying.'

'We need to make them run faster. I need a horse!' the Rambling Pommy declared in the manner of a man with a mission.

'You need to howl like a pack of dingoes on the bloodtrail,' Kelpie put in vigorously. He hooked his thumbs in his belt, bushman style, arched his back and let rip a blood-curdling howl that turned the Pom pale and set the Kelpie dogs to howling mournfully in response. 'You've got to scare 'em shitless! Bore it up 'em – so they know it's FLY or DIE!'

With his blood running hot, the Pom confronted Jeff. 'Boss, can I borrow your saddle horse tomorrow morning?'

The grazier was caught short. He felt the joke was getting out of hand and only with difficulty maintained a straight face. Sparring for time, he rubbed his chin and mulled a moment before replying, 'Well – er, that's fine, but

I doubt any of my mounts would be hefty enough to carry you. You're a big fella. What do you tip the scales at, son?'

'I was fifteen stone when I did army training; and sixteen stone when I played front-row second division last season. I'm carrying a spot of lard, so I might be seventeen stone now. I stand six feet and one inch in my socks.'

Jeff was still hesitating, when spurred by desperation and destiny Cyril added pointlessly, 'I'll turn twenty-one years of age on February the twenty-first, Boss. My mum wants me home for that occasion. I'm joining the army then.'

'I'm sure she does,' said Jeff, still clamping his jaws to prevent a mirth explosion. 'Well, ah, we've established that your weight might be a problem, but first and foremost – can you ride?'

'Ride!' the Rambling Pommy exclaimed, stunned momentarily by what seemed a stupid question. I've done farm work all my life! I can drive any tractor you like to name. I've got an A-one in animal husbandry and a B-two in large acreage agriculture . . .'

'Well, ah, I must say that's impressive.'

Kelpie's lofty form loomed up beside Jeff. 'We can't disappoint the lad, Jeff,' he said solemnly. 'He's come halfway around the world to see emus fly. In the mornin' I'll saddle Maisie, that old half-draught mare o' mine. She could pack a piano.'

Everyone who owned a camera brought it to the shed. They were mostly Box Brownies, but Kelpie, who sold photos to the *North Queensland Register* and other country newspapers, sported a 35mm job with a zoom lens, while Jeff carried his wife's 8mm movie camera.

On cue, the emus fossicked for tidbits along the bore drain. The Rambling Pommy was so excited he skipped smoko, swung into the saddle and jogged towards the target, with Fritzy trotting beside and the cameramen close behind.

The Pom's familiarity with the saddle surprised the onlookers, who were expecting him to go overboard quick smart, for Maisie was a sharp turner: she had served time droving and camp-drafting cattle, and was indignant at her rider's poundage and being asked to muster emus. Like Don Quixote and Sancho Panza they made a ludicrous combination, the big mare dutifully rolling into a jog and the hefty Yorkshire lad posting in the saddle while faithful Fritzy ran beside.

Coming onto the emus, the Pom and Fritzy yelled and howled with the zest of true believers. Their enthusiasm infected the apostates, and even Kelpie broke into a jog to keep up with the shouting, laughing and yipping chasers as he positioned for photographs.

The emus swung into a swift trot towards the homestead, and the galloping goats and kids joined them. The mare followed closely with the Pom shouting and waving his hat, while the foot-sloggers snapped their photos and laughed their way back to the shed for a postponed smoko.

24

THE WHITE GODDESS

Despite his misgivings Jeff was enjoying the joke until he glanced out the wool-room window and saw a car entering the homestead grounds 400 yards away. 'Holy smoke!' he exclaimed. 'The wife is home! I wasn't expecting her till the weekend.' He swallowed his tea, grabbed his hat and said, 'C'mon, Kelpie. We'll leave the branding till later. I'll take the ute and the dogs, you saddle up, and we'll go and give Silent a hand to muster those two-tooth wethers.'

Kelpie got to his feet with uncharacteristic speed. 'Good thinkin', Jeff!' he said. Turning at the exit he offered a piece of missing information. 'You fellas! Jane raised those three birds since they was chicks. My guess is she'll take a dim view of a maniac she wouldn't know from Darcy Dugan chasin' them around – on *my horse*. I'll see yer later.'

A few minutes after shearing resumed, pounding hooves and a shout of 'Whoa!' announced a furious Yorkshire lad's return. Raging through the loading doorway, he cut loose: 'You're a mob of bloody gormless idiots, that's what you are. All matey with big Cyril to his face and taking the mick behind his back. And you've made me a fool in front of the Mrs. "Emus *can't* fly, you silly boy," says she.'

He clumped half a dozen steps, swinging his head like an angry fighting bull in search of his tormentors. 'Where's that Kelpie fooker? *Silly boy*, huh! Ah'll knock his bloody block off!'

Sitting on a wool bale, I primed myself for an evasive roll off the far side, but was caught by the explosive speed of the front-rower's charge. My twelve stone was folded as easily as a football into a huge right arm, carried three paces and hurled bouncing across the flagstone floor.

The shearers had their heads down at work, while the rousies watched from the far side of the wool bins as Dick confronted the berserker. 'Hold it! Hold it! Settle down!' the overseer commanded. At arm's length the giant dwarfed the overseer. 'Shut up and step aside before I lose my rag and snot you,' Cyril snarled, shoving a mauley as big as a leg of lamb under Dick's nose.

I got swiftly back on my feet, deciding on inconspicuous retreat, but when I saw an old mate about to be steam-rolled by a raging giant, I slipped up behind the Pom, aiming to king him with a right to the jaw. *It better be a good'un*, I thought, badly shaken by my bouncing on the flagstone floor. I was relieved to hear Dick call on his army experience to take command: 'Attention, soldier! Attention!' he roared. 'Another word and you will earn three hours on the parade ground and confinement to barracks. UNDERSTAND?'

Cyril had straightened his posture and was halfway through a salute before he realised what he was doing. He shook his head while muttering a volley of hostile Yorkshire profanity, but the impetus of his rage was broken.

Dick stared him down and then grinned. 'You're on duty in five minutes, soldier. Grab a cuppa and some smoko and then go to your post.'

A few minutes later I was by a wool-room window washing grit and gore from my abrasions in the hand basin, when a dusty Falcon sedan parked in the shade of a tree near the wool room. Having passed most of my working life in a womanless world of bush-camps, I enjoyed a serendipitous vision as the driver, a good-looker I reckoned to be in her late thirties, slipped swiftly from behind the wheel. She was wearing a floral summer frock and sandals. *The boss's wife*, I assumed, as the woman stood for a moment taking in the scene, smoothing down her frock and brushing back some vagrant blonde hairs.

Apparently satisfied with her appearance, she clamped on an aggressive frown which threatened retribution, and swung into a purposeful march towards the wool room. Her striding attitude changed my prospect, morphing her shapely calves and sandalled feet into a Roman centurion's muscular foot-slog. *I fear no man and very few women*, I had often quipped, and this furious femme was definitely a candidate for the latter category. I ducked into a bin, yelled 'Ducks on the pond,' a traditional warning to show respect to women entering the shed, and began arming wool, leaving Dick to face the Wagnerian music.

'Good morning, ma'am. Welcome home – and it's a pleasure to see you.'

'What the devil is this formality bit? The name is Jane – as you well know!'

'Well . . . Ah, yes . . . Jane. Can I be of assistance?'

'My God, Dick! What is this? My shearing shed, or Tom Fool's restaurant? *Can I be of assistance?* What's next – a wine list? Cut the bullshit! You know why I'm here. Galloping my emus and milking goats all over the place. *Flying emus!* You ought to be ashamed of yourselves. And that poor lad, Cyril – comes all the way from Yorkshire to see our lovely country, only to be made a fool of by a mob of pathetic morons. Your men are way out of control, Dick. *Way out of control!*'

'It was only a joke that got a bit over-cooked, Jane,' Dick said, his confidence firming. 'I promise you it's all over now. All is well. A good bright wool clip; and I'm counting out over a thousand shornies a day.'

'Don't try to change the subject, Dick. Like hell it's all over! Where's Jeff? And where is that gangling fool, Kelpie? He was a kid when he first came here fifteen years ago – and he's grown backwards every year since. This shenanigan has got his brand all over it!'

Dick was one of those men who enjoyed any exchange with a woman at long and medium range. At distance he was a dab operator, confident of his ability to oil rough waters and relieve and amuse dissatisfied moods with empathy and laughter; it was only when he was tested for prolonged periods at close range – through marriage or de facto relationships – that he discovered the companionship of 'The Female of the Species' to be unendurable.

Within minutes Dick and Jane were enjoying each other's company, laughing and chatting. They walked around to the shearing board, where her womanly presence instantly put a hold on the flow of banter and profanity. She nodded to the workers when she caught their eye, and chatted breezily to Ritzy – an old acquaintance – at the pen gate, before singling out a blushing but delighted Cyril for friendly attention. Observing Ho Chi Minh hitching his dungarees around to hide the gap left by his missing fly buttons, I recalled Lawson's take on ladies in the shearing shed:

> *'The ladies are coming,' the super says*
> *To the shearers sweltering there,*
> *And 'the ladies' means in the shearing shed:*
> *'Don't cut 'em too bad. Don't swear.'*
> *The ghost of a pause in the shed's rough heart,*
> *And lower is bowed each head;*
> *And nothing is heard, save a whispered word*
> *And the roar of the shearing shed.*

As she left, the boss's wife called from the door, 'Dick! Tell Kelpie that I want to see him quick smart! I'll have his hide for boot laces! And I suspect Jeff was in on this, too.' I considered making her acquaintance, but caution overruled desire and I didn't emerge from the sheltering wool bin until I heard the Falcon drive away.

25

SNOWBALL, DICK AND THE EAGLE

After knock-off time, a shower and a couple of Fourex longnecks restored the Rambling Pommy to his boisterous, jovial self. The emu affair, reviewed from various angles, inspired big mobs of mirth – me copping more chaff than the Rambling Pommy, who challenged me to a return wrestling bout: 'Yorkshire rules, you little fooker. No holds barred, head-butts and forearm jolts allowed.'

'My mother never reared a squib, Rambler; but the conditions are that you first ride a bucking emu to a standstill, use only one arm, and get down to my weight.'

On Friday morning Ho Chi was again irascible and on the prod. Over dinner The Eagle called out, 'Who'll be here for the weekend? The old Wedge-tail will be here to put on a feed, but the more of you he ain't got to muster grub for the better he'll like it.'

'I've tallied 'em up, Eagle,' Dick announced. 'There's only you and me and the presser.'

'I think I'll stay,' Ho Chi announced insincerely.

'Well, make up your bloody mind, Chinaman,' Dick advised. 'That's not what you told me an hour ago. The Eagle wants to know.'

Ritzy stirred the possum by offering to help out. 'You'd better come to Hughenden with me for the weekend, Ho Chi. Plenty of giggle-weed and girls. In fact I can point you in the direction of a long-haired mate called Dancing Daisy. She's about your style: a bit rough around the edges, but they don't call her "Smoke 'n Poke" for nothing.' He joined in the laughter and added, 'Take care of that little lady, Ralph, and she'll make you a good wife – just what you've always wanted.'

Ho Chi wasn't laughing. 'You married blokes give me the shits. You tell everyone you're having a ball – but you're really sulking in the marriage slammer, and making the best of a bad situation. Human bondage, they call it.' He quoted, '"Marriage is an institution which turns life's greatest pleasure into a duty."'

The Eagle said, 'Make up your bloody mind, runt.'

'Why don't *you* go to town, you great bag of lard. Then we can all jump out of the play-pen and fly kites.'

The Eagle had been copping more than his share of Ho Chi's bile. He wasn't armed with the sharp repartee necessary to spar with the shearer, so he usually laughed the jibes off or chose silence. Now he counted to ten before replying, 'I'm putting a bank together – I haven't had a punt for three months. I was planning to go to the Cup again, but I gave it a miss. Still, I'll be in Melbourne over

Christmas to see my daughter and her husband and my grandkids.'

'The CUP!' Ho Chi said sarcastically. 'A likely story! And I suppose Bart Cummings invites you to drinks at the VJC members' bar?'

Getting no bite, he continued to cast: 'Ha ha, that's rich! Daughter and grandkids! How would you get a wife? You make Frankenstein look like Clark Gable.'

A few thumping steps and The Eagle swooped upon his persecutor. His face was red and convulsing with anger. His huge hands encircled Ho Chi's puny throat and he snarled, 'Shut up, you runt! You don't know what you're talking about. Shut up or I'll bloody well choke you.'

Ho Chi gasped and gargled until the Rambling Pommy rose swiftly and firmly forced the babbler back, while Dick commanded, 'Settle down! Easy! Take it easy.'

Breathing heavily, almost sobbing, The Eagle murmured, 'I had a wife . . .' and went to his bedroom.

I asked Dick as they walked back to the shed, 'What do yer reckon stirred the babbler?'

'Well, he's a cook – and it don't take much to get a babbler offside, especially putting up with a little shit like the Chinaman. From what I gather he's got his heart set on seeing his daughter and grandkids. He's never met the kids and hasn't seen his girl since she was sixteen. His wife passed on years ago.'

It was customary for me to head for the kitchen at a quarter to three to assist the cook to carry the smoko to the shed. Jogging past the overhead tank I called politely without looking, 'Greetings, oh Lord of the Flies.' I heard Snowball grunt acknowledgement.

Entering the kitchen I called breezily, 'Righteo, Chef. Let's go!' Getting no answer, I fired a volley of cooees, and did a check of the cook's bedroom and the toilets, but The Eagle was absent.

Remembering Happy Jack twenty-odd years earlier at Kahmoo station, a chill of horror halted me for a moment before I opened the meat-house door. To my immense relief The Eagle wasn't sitting in there, a shotgun between his knees, bloody remnants of his skull and brains scattered about the ceiling and gauze fly-proofing. The room was empty, its white walls and cement floor scrubbed clean.

On the kitchen table sat a tray of sandwiches covered with a damp tea towel, and another of cakes and biscuits. In the range oven were a dozen hot sausage-rolls; water was boiling and tea was in the urn. I made the tea, balanced two trays on one arm, picked up the urn and headed for the shed.

Smoko was over and shearing underway before I said to Dick, 'I'm a bit concerned about the babbler. I couldn't find him before smoko, and he hasn't collected the smoko gear.'

'Why worry, Presser? He's big and ugly enough to look after himself,' Dick replied, and turned back to grinding combs and cutters.

'There's a Bundy rum bottle on the kitchen table – with only a nip or two left. I reckon it's one of yours.'

Dick swore expressively and we trotted to the kitchen. We searched the huts and cooeed through a nearby patch of scrub. 'Stupid old bastard; he's probably gone walkabout and choked under a bush,' Dick muttered. 'We'll go back to the shed and give him half an hour. If he doesn't show you go to the homestead. Maybe he's gone there to ring up or something. If he doesn't turn up we'll knock off shearing and start searching.'

As we strode past the tank, Dick stopped short. 'Whoa! Whoa there, boy! You need bloody spectacles, Presser. Cast an eye!' The two large bodies, lying down side by side, their whiteness camouflaged with brown slush, looked almost identical. The big boar reacted to a prod from Dick's boot by standing and squealing in protest. 'I reckon that one is the pig,' he commented dryly. We pulled The Eagle, swearing and stammering, into a sitting position. 'He's alive, thank God!' I said.

'Yeah, but he don't smell like he is.'

'Looks as if he fed his old mate, then tripped over the scrap bucket,' I observed, indicating a badly bent aluminium bucket.

Dick grinned. 'An eagle as pissed as a parrot! Spare me days and give me strength! I thought I'd seen it all!' He dried his hands on his jeans and the tail of his work shirt and rolled a smoke. Then, resting against the tank stand, he inhaled a relaxing drawback. 'I'll wager a packet of Drum to a pinch of sheep shit that The Eagle don't tell the grandkids about this caper.'

His voice took on a steely edge. 'And we won't be saying anything, either, Presser. It would be dishonourable. We can't make a laughing stock of a man who fought for his country. We'll have to watch him for a day or two – when he thinks this through the old bloke's self-respect will be lower than a black snake's shadow.'

I was taken by surprise: already I'd been thinking of the laughs the yarn would raise. I felt a measure of shame that I hadn't had the decency and gumption to think twice.

Old words flashed into my mind: *honourable, noble, loyal* ... They were in common usage around the bush when I was growing up. Now only a few of the older men and women employed them, so they were rarely heard. Had such words all but vanished from the bush vocabulary?

I looked up and met Dick's earnest gaze with renewed respect. 'I'm with you, Dick, a *noble* consideration. You're bloody right.'

'We'll walk him to the washhouse and shove him under the shower – clothes and pig shit and all, and then throw him onto his bunk. I reckon he'll choke for hours. You can knock off early and keep an eye on him till the boys get away. Then tomorrow morning we might take him to Hughenden with us, have a few friendly drinks and a feed to cheer him up, and bring him home before he gets on the punt.'

The Eagle began mumbling between sobs, 'Got to see my daughter. I love my girl, she's my life ... Bitch – she never wrote. The kids are named after my darling wife. You don't think I can remember their names, Dick?' Giggling between tears, he rambled on. 'Old Wedge-tail is smarter than you think ... Dick, my good friend, Dick – best boss-of-the-board in Queensland. Help me up ... Gotta get up and put a feed on, it's getting late.'

'Shut up, Eagle!' Dick snapped. 'Fighting drunks I can handle – crying drunks I won't wear.'

The boar had remained an interested spectator, standing in the wallow a few yards away; now he was fixing us with piggy eyes, mixing guttural grunts with excited squeals.

I said, 'I reckon Snowball takes a dim view of the way we manhandled his mate.'

Out of habit Dick's farmer's eye had been admiring the boar's lines. 'You wouldn't see a better animal at the Royal Easter Show, Presser – pedigreed without a doubt. I reckon Snowball could win a ribbon in Sydney if I dolled him up.'

Snowball squealed porcine approval. He rose, turned a disdainful rear end on The Eagle, and swaggered towards the bore drain.

'Too right, old fellow!' Dick congratulated. 'A chap has got to maintain his standards.' He chuckled – and quoted:

You can tell a man who boozes
By the company that he chooses;
And the pig got up and slowly walked away.

I laughed aloud at the time-worn quote. 'Too right, Dick! Dad often quoted those very lines. It's a pity you didn't meet him. He was a good horseman, like yourself. You'd have been good mates.' My grin faded to a wistful smile as images of my father, who had died of heart failure three years earlier, occupied my mind.

I remembered Dad, light of foot, swinging into the saddle of a prancing filly; Dad on the shearing board, grimacing as he stood to ease his aching back, then bending to the task of earning for his family. And I thought about that anxious father, meeting his physically fragile son on the Cunnamulla railway platform more than thirty-five years before. Back then I had been his shiralee: his burden. Little did either of us realise that my twenty-four-hour train trip heralded the beginning of a journey into adulthood, and to health and fulfilment. The Boy would grow into the Presser on a unique voyage through a rugged but rewarding life in the outback wool industry.

ACKNOWLEDGEMENTS

Three cooees for Robert Macklin, still a bushie at heart, who urged me to write a book about 'those fabulous Babblers of the Bush', and provided endless encouragement from go to whoa. Thanks to Margaret Kennedy, who enthusiastically convinced me that I had produced a publishable work, and swiftly placed the book with Random House. Thanks also to all the team at Random House, who professionally 'pressed up' *Wool Away, Boy!*, especially Tamika Wood for publicity; and a huge thank you to commissioning editor Sophie Ambrose whose patience and care and warm personal touch permeated her wise advice and direction chapter by chapter. Last but not least, *Wool Away, Boy!* is a tribute to our wonderful wool industry, and the shearers, rousies, babblers, pressers, classers and overseers who combined to harvest the clip when Australia 'rode on the sheep's back'. They enriched my life with the values of hard work, mateship and laughter.

GLOSSARY

axle-buster A road bump so severe that it destroys axles

babbler An abbreviation of 'babbling brook', rhyming slang for 'cook'

backing a tail for a dollar Common idiom for 'five bob' (five shillings)

bag boots Bag boots were a type of slipper generations of shearers wore. They cut a piece of jute from woolpacks using hand shears, shaped it and then sewed it with a large bag-sewing needle. The jute was clustered into a knob over the toes for protection, and laced with twine. In the late 1960s nylon packs, unsuitable for boots, replaced jute

bait-layer An unflattering term for a cook. You risked death eating a bait-layer's tucker

GLOSSARY

barrowing A shedhand learning to shear

Beetle A Volkswagen sedan

billy lids Rhyming slang for 'kids'

blackened wool Wool harvested by 'blacklegs' (strike breakers/'scabs'), which the transport unions refused to handle

blue heeler An Australian-bred cattle dog, also known as 'bluey'

bodgie Falsify

bog-eyes or 'bogis' A shearer's mechanical hand-piece

bomb Reprimand

Bosca of the Bog-eye Terminology shearers used to introduce themselves, especially to those they thought snobbish

brace To confront, demanding explanation

brolga A large crane which performs an elaborate dance

brownie A basic fruit cake flavoured with cocoa (sometimes sans the fruit and cocoa)

bunyip A mythical large, horse-headed swamp dweller reported by early white settlers

butts Wool-packs containing wool from the board collected by hand. When empty they were 'packs', became butts when containing wool, and bales when pressed. Three full butts made a pressed bale

cackle-nut farm Poultry farm

carbide light Gas lamp

cauliflower ear An ear permanently thickened, usually a trademark of boxers

chuck a grenade Reprimand. Other common slang terms were 'bomb' and 'chip'

cobbler The toughest sheep in the pen

colours Discarded chips of opals

Creek An abbreviation for Julia Creek

Cup The Melbourne Cup

Curry Abbreviation for Cloncurry

cut-out When the last sheep is shorn, the last bale pressed and the men paid off

dawnies Alcoholics in quest of an early drink

dero Short for 'derelict'; a pejorative term for a homeless person

deuceing Shearing 200 sheep

dobber An informer

doggers Professional dingo hunters

Downs tiger A huge, venomous king brown snake, also known as a mulga snake

dreadnought Someone capable of shearing 300 sheep in a day – a rare tally

drongo A person who is as thick as a brick

ducks on the pond Traditionally called out as a warning to show respect to women entering the shed

GLOSSARY

dunnies Toilets

frog and toad Rhyming slang for 'road'

gentleman of the long tube Shearers would jokingly call themselves 'gentlemen of the long tube' when trying to upgrade their social standing

gigging Teasing, also known as 'chiacking' or 'taking the mick'

guard A reliable and favoured employee

Hill (The Hill) Broken Hill

hit the toe Depart hurriedly

hoop Idiom for jockey

ice chest A frame covered in hessian, cooled by water dripping in a shady, breezy spot

Isa (The Isa) Mount Isa

jackaroo A young man gaining practical experience on a sheep or cattle station

Joe Blake Rhyming slang for 'snake'

Kerosene Bouquet Coarse laundry soap

khyber An abbreviation of 'Khyber Pass', rhyming slang for 'arse'

kip A small board from which pennies were thrown when playing Two-Up (manipulative fingers were not trusted)

knuckle sandwich A punch

Koertz A common make of wool press. Most good pressers considered it inferior to the Ferrier

locks-butt A wool-pack containing oddments

London to a brick Those in no doubt of winning will bet London to win one brick. The expression was made famous by Ken Howard, the doyen of Sydney race commentators in the 1950s and 1960s

long paddock Idiom for the stock route: a continental maze of interconnecting tracks which allow the movement of stock the length and breadth of Australia

longneck Traditional 26 oz brown bottles of beer. 'Big bottles' and 'king browns' are other idioms

mauley (or maulie) A fist, especially a fighter's fist

min min A mysterious light seen occasionally in outback Queensland

monkey board A square board fitting on the top of a press tamped tightly with wool. The 'monkey' compresses two boxes of wool into one bale when the presser swings on the lever

mulga wire Long distance word of mouth; also known as the 'bush telegraph'

Mum The title given to a motherly working woman by lonely lads far from home

on the board The place within the shed where shearing actually takes place

GLOSSARY

picker-up The person who picks up fleeces on the board and throws them on the wool-roller's table. The knack is to give fleeces flight so they descend spreading evenly on the table. Picker-ups were usually teenagers. A capable 'boy' could pick up 1000 or more fleeces a day for six shearers

piece-picker A rouseabout who sorts the lower grades of wool, removed by the wool rollers

pitch and toss Rhyming slang for 'boss'

Pix A common exclamation of doubt, referring to a popular magazine

plum jams Rhyming slang for 'lambs'

pointers People who point the finger at others behind their back

porky Short for 'pork pie', rhyming slang for 'lie'

port A suitcase

possie A good position

potch Low-grade opals of no commercial value, usually discarded by early miners

press-up A presser always finished last, after he had 'pressed-up' the last of the wool

provo Slang for a member of the police

Q fever A debilitating virus carried by sheep and transmissible to humans

Queenslander A house on high stumps with wide verandahs, built for the climate

Reach (The Reach) Longreach

ring Collar the fattest cheque. In smaller sheds the fastest shearer would 'ring' the shed; in larger sheds where there were eight to ten shearers the contract cook or presser might get paid the most

ringer A mounted station hand; also the fastest shearer in a shearing shed

rouseabout A shedhand

rousie Abbreviation for 'rouseabout'

scratch-pulling A game where opponents sat on the floor, with the soles of their feet pressed together and arms outstretched, holding a broom handle between them. The aim was to pull the other up

sheep cocky Originally a small land-holder; later used to identify any owner of sheep, regardless of property size

shiralee A swagman's bundle; a burden of responsibility

skirt Wool rollers would 'skirt' (or 'roll') the rough wool off the fleece before it went to the classer's table

snag An obnoxious troublemaker

snagger A rough, slow shearer

snobs Sheep that are tough to shear

soul-case Slang for 'body'

Spaniard Common workers' idiom derived from 'Manuel [manual] labour'

speared Sacked

stubby A half-sized bottle of beer

stud rouseabout A learner classer/overseer

GLOSSARY

tenner Ten-pound note

tinnies Tins of beer, half the quantity of a longneck

Tojo Bush idiom for a Toyota LandCruiser

tomahawking Shearing so rough and bloody it might have been done with a tomahawk

troppo From 'tropical'. Soldiers serving too long in tropical conditions were said to go troppo (crazy)

tucker Food

tumbling-tommy A cook's large rubbish bin on two wheels, parked outside the kitchen door

two-tooth A two-tooth wether is a sheep with its first adult teeth

Two-Up A game in which people bet on whether two coins (tradionally pennies) will land Heads or Tails. The coins must be tossed cleanly and spin or a re-throw is called. Three coins could be used to speed up the game, bets being paid on a majority of two Heads or Tails

wet-voter When it rains and sheep get wet shearers take a majority vote: a dry or even vote means you carry on, a wet vote means knock off

wool pressing The heaviest work in the shed. The wool presser compresses the wool into a wool press making 'bales'

wowser A killjoy

yakka Hard manual work

Yowies Australia's version of the Himalayan Abominable Snowman and America's Big Foot

AUTHOR'S NOTES

Chapter 1

p. 6 'It was only the powerful young wool presser, Jim, who was kindly and understanding of a lad's needs.' Not many men were robust enough to take on the lever. A few pressers also worked cane-cutting and wheat lumping. These three contract jobs were recognised as the toughest occupations in the bush.

Chapter 2

p. 20 '. . . it was a matter of pride among professional lever-men to keep the wool away within the bell hours'. Nearly all employees worked Pastoral Industry Award 'bell' hours. There were four two-hour 'runs', making eight hours per day. The only exception was the contract presser, who was allowed to work any time between the 7.30am Monday

AUTHOR'S NOTES

starting bell and the 5.30pm Friday knock-off bell, because the quantity of wool came unevenly. For example, while ten shearers might shear only fifteen bales a day off lambs, meaning the presser had to wait while the bales built up, they might manage forty or fifty bales a day off full-grown sheep and the presser might have to work outside bell hours to keep up.

p. 20 'Muscle-sore and dog-tired I plodded to the shed each night after tea with a carbide light and a proud can-and-will determination.' Many workers joked that carbide lights were an explosive gas device masquerading as a lamp. In fact, if serviced correctly they gave a bright steady light until a surge of gas would ignite, exploding into frightening flashes of flame and loud bangs – which would send the uninitiated scampering.

p. 30 'He's headed 'em twice but he can't do it thrice . . .' Often quoted folklore lines.

p. 31 'I love you in your negligee . . .' Origin unknown. Certainly familiar to Tommies and Diggers in World War II and often heard among bush workers thereafter.

p. 31 'When he moved on to Banjo Paterson's iconic ballad, with a few boozy voices harmonising to his mouth organ, Karl queried, "What is this 'Waltzing Matilda'?"' Banjo Paterson was an icon of Australian bush poetry and seminal national identity.

p. 32 '. . . he whipped into the fray like a southpaw whirlwind, weaved inside like a professional, and hammered me into a circling retreat.' Right-handers learn to box with the left leg and hand (paw) forward. Their stance is called orthodox. Left-handers box with the right leg and hand forward, called southpaw.

p. 32 'Cop that young 'Arry!' The phrase comes from comedy star Roy Rene's sketches.

p. 33 'Anyway, ridgy-didge, where did you learn to box like Jimmy Carruthers?' Jimmy Carruthers was Australia's first universally recognised world boxing champion.

Chapter 3
p. 42 'Talk about a Clive-bloody-Churchill tackle, Bluey! You're a little ripper!' Clive Churchill was a legendary rugby league fullback who played for New South Wales and was captain of the Kangaroos.

Chapter 4
p. 55 'I reckoned the ringer looked old and brown enough to be an original de Satge offspring . . .' Oscar John de Satge was born in England in 1836. He arrived in Melbourne in 1851 fresh from Rugby School, and became a clerk on the goldfields before heading to Queensland to become a drover, pioneer pastoralist and influential politician. In 1884 he took up Carandotta with two business partners, and entered into a liaison with an Indigenous woman who gave birth to two sons and a daughter. After drought in the early 1890s killed 90,000 sheep and 15,000 cattle on Carandotta, de Satge returned to England, where he married Beatrice Elizabeth Fletcher. The de Satge name lives on in the north-west through his descendants. A granddaughter, Ruby de Satge,

AUTHOR'S NOTES

became famous for droving cattle throughout the north-west and the Northern Territory.

Chapter 6
p. 76 'Seeing a chance I couldn't let slip, I bunged on a Dad and Dave drawl . . .' Dad and Dave were humorous bush battlers in the popular radio serial of the same name, based on Steele Rudd's (aka Arthur Hoey Davis) stories.

p. 79 'Who does he think he is? Herb Elliott?' Herb Elliott was an Olympic champion and world record holder for 1500 metres and the mile. He was a household name.

Chapter 7
p. 82 'You're faster than Wild Bill Hickok, Wide Awake!' Wild Bill Hickok was a household name in the 1950s, courtesy of his role as a hero in Hollywood westerns. The original Wild Bill, a retired town marshall and multiple man-killer, was murdered in a poker game in Deadwood in 1876, while holding aces and eights – still known as 'the dead man's hand'.

p. 84 'The song I'm going to sing you will not detain you long . . .' This is an old Australian folk song called 'The Station Cook', which the American folk singer Burl Ives revived in 1952.

p. 98 'Darcy Dugan wears clobber like that when he holds up a bank.' Darcy Dugan was a notorious bank robber and jail escapologist.

Chapter 8

p. 109 'No thank you, Boss, I'd rather not . . .' An oft-repeated quote from 'The Stockman's Tale' (Anon.).

p. 109 'You can talk of your whisky, talk of your beer . . .' From 'Billy of Tea', a traditional song.

Chapter 9

p. 116 'Using fencing pliers he wired the exhaust pipe to the chassis of the Morris Minor with a Cobb and Co hitch.' Named after the Cobb & Co coach line, this was the most common, secure and simple wire hitch.

Chapter 12

p. 151 'Are you going to invest in the Bill Waterhouse retirement fund?' Bill Waterhouse was the doyen of Sydney bookies. A colourful character and household name, he frequently made the headlines.

Chapter 13

p. 158 'A fight draws spectators like a bush killing block draws flies and meat ants.' All sheep stations and shearers' quarters had an adjacent rectangle of concrete called a killing block and a gallows for suspending slaughtered sheep to be bled and skinned.

p. 158 'He wore grey slacks and shined Julius Marlow shoes.' Julius Marlows were an expensive brand of shoes.

p. 164 '"It's not the Black Arabian Pox," he said quietly.' The Black Arabian Pox was supposedly an incurable venereal disease brought back by soldiers from the Middle East. The yarn was used by old hands to terrorise young men: 'You go blind, your cock falls off, and you're dead in three weeks.'

Chapter 14
p. 174 'Too correct, mate! And I was the picker-up when Jackie Howe tallied three hundred and twenty-one at Alice Downs in 1892.' Jackie Howe was a legendary Australian shearing champion, and stalwart of Labor politics.

Chapter 15
p. 204 'I'm no Don Athaldo . . .' Don Athaldo was a famous Australian strongman. One of his most spectacular demonstrations of strength was carrying a pony up a ladder. He also sold body-building courses.

Chapter 21
p. 260 '. . . a double for Dolly Dyer!' Dolly Dyer was a TV personality and the wife of the famous performer Bob Dyer.

Chapter 22
p. 269 'Around Hughenden and in the north-west, where I worked with Ritzy, they call a king brown or mulga snake a Downs tiger; but around Longreach it seems a Collett's snake is often dubbed a Downs tiger. Take yer pick!' A Collett's snake is often mistaken for a Downs tiger. Similar in appearance, a Collett's snake can grow to seven feet in length, king browns to nine feet.

p. 270 'Yer kids will miss out on Christmas if you presser-starvers don't fill the comb and push a bit harder.' An essential part of good shearing is 'filling the comb'; that is, using the full width of the comb to cut wool at each stroke, known as a 'blow'. The more you fill the comb the less blows you have to make.

Chapter 24
p. 280 The chapter title 'The White Goddess' is a reference to the pagan goddess inspired by the moon and lunar phases. Identified in Robert Graves fascinating book of the same name.